Science and Social Context

Science and Social Context

The Regulation of Recombinant Bovine Growth Hormone in North America

LISA NICOLE MILLS

McGill-Queen's University Press
Montreal & Kingston · London · Ithaca

ISBN 0-7735-2374-X (cloth)
ISBN 0-7735-2375-8 (paper)

Legal deposit third quarter 2002
Bibliothèque nationale du Québec

Printed in Canada on acid-free paper that is 100% ancient forest free (100% post-consumer recycled) and processed chlorine free.

This book has been published with the help of a grant from the Humanities and Social Sciences Federation of Canada, using funds provided by the Social Sciences and Humanities Research Council of Canada.

McGill-Queen's University Press acknowledges the financial support of the Government of Canada through the Book Publishing Industry Development Program (BPIDP) for its publishing activities. We also acknowledge the support of the Canada Council for the Arts for our publishing program.

National Library of Canada Cataloguing in Publication Data

Mills, Lisa Nicole, 1967–
Science and social context : the regulation of recombinant bovine growth hormone in North America
Includes bibliographical references.
ISBN 0-7735-2374-X (bound).—ISBN 0-7735-2375-8 (pbk.)
1. Recombinant bovine somatotropin—Government policy—United States. 2. Recombinant bovine somatotropin—Government policy—Canada. 3. Recombinant bovine somatotropin—Economic Aspects. 4. Monsanto Company.
4. Monsanto Company. I. Title.
SF98.S65M54 2001 636.2'0892405 C2002-900052-1

Typeset in 10/12 Sabon by True to Type

Contents

Acknowledgments

This book is based on my PHD dissertation, and I would like to thank the members of my dissertation committee for their support and encouragement, especially my supervisor Professor David Wolfe and Professor Liora Salter.

I also wish to acknowledge all of the individuals whom I interviewed as part of the research, particularly officials at the U.S. Food and Drug Administration, who were generous with their time both during the interviews and in subsequent communications. The following institutions and individuals also made important contributions to the completion of this project: the Toronto Food Policy Council gave me access to its library; Dr Alison Weir assisted me with statistics; Dr Magi Abdul-Masih assisted me with concepts in biochemistry; and Patricia Fleck gave valuable assistance with typing the bibliography and checking interview quotes. The manuscript has benefitted greatly from the comments of two anonymous reviewers. All errors and omissions are, of course, my own.

The dissertation was written with financial support from the University of Toronto Open Fellowship and a School of Graduate Studies Travel Grant.

I would like to thank the staff at McGill-Queen's University Press, especially the executive director, Philip Cercone, for his support, Ron Curtis for excellent copyediting, and Joan McGilvray and Brenda Prince for co-ordinating the editorial process.

I also wish to thank my family and friends for their support, particularly the Loretto Sisters, Extended Community, Leadership Team and staff; Catherine and Peggy Lathwell; Dr Leslie Crawford; Dr Katie Isbester; Dr Mary-Beth Raddon; Dr Joe Murray; and Deborah Clipperton.

Chronology

UNITED STATES

1982 The first study of rbGH is conducted at Cornell University.

1985 The FDA permits the release of meat and milk from test herds into the milk supply after reviewers decide that food from treated animals is safe for human consumption.

1986 The House of Representatives Committee on Agriculture opens hearings on the potential impact of rbGH. Anti-biotech activist and head of the Foundation on Economic Trends, Jeremy Rifkin, represents a coalition of organizations opposed to the drug. Representatives from each of the four companies speak. The Co-ordinated Framework for the Regulation of Biotechnology Products, which specifies that biotechnology products and processes will be regulated under existing statutes, is announced.

1987 Rifkin petitions the FDA to consider the economic consequences of the introduction of the drug; the FDA rejects the petition, on the grounds that economic considerations are outside its mandate.

1989 A summary of Monsanto's animal toxicity testing is leaked to Dr Samuel Epstein of the School of Public Health, University of Illinois. Epstein argues that increased levels of a growth factor in milk could lead to a higher risk of breast cancer in the general population. In response, Rifkin's Foundation on Economic Trends petitions the FDA to ban the sale of milk and meat from cows in test herds. An FDA reviewer, veterinarian Richard Burroughs, is fired by the FDA. He alleges that he was dismissed because of his concerns about the animal health data.

1990 FDA scientists Judith Juskevich and Greg Guyer publish a summary of the agency's human health evaluation in *Science*. A National Institutes of Health (NIH) panel reviews the data on human and animal health and concludes that the drug is safe; however, not all the animal health data are available to it.

1990 The *Journal of the American Medical Association* and the *Journal of the American Pediatric Association* endorse the safety of the drug.

1991 Congressional representatives ask the General Accounting Office to investigate FDA actions regarding drug trials at the University of Vermont, and they ask the inspector-general of the Department of Health and Human Services to investigate Monsanto's pre-approval promotion of the drug.

1992 The GAO investigates the FDA's human health evaluation and advises that it should be broadened to examine the indirect risk to human health from increased levels of mastitis in cows.

1993 The FDA's Veterinary Medicine Advisory Committee meets to consider the issue of an indirect risk to human health from the increased use of antibiotics to treat mastitis in herds and agrees with the FDA's conclusion that such a risk is "manageable." The FDA's Food Advisory Committee meets to consider labelling milk from treated cows and decides against mandatory labelling, because the milk is not significantly different. However, the FDA does allow for voluntary labelling of milk from untreated animals, provided that the label is accompanied with a disclaimer noting that there is no significant difference between the two kinds of milk. The FDA approves the drug for sale, subject to a two-year post-approval monitoring program (PAMP). Congress enacts a ninety-day moratorium on the product, during which the Office of Management and Budget (OMB) reviews its likely impact. The OMB concludes that the approval of the drug could enhance U.S. leadership in biotechnology.

1994 February. Monsanto's product Posilac goes on sale in the United States.

The *Los Angeles Times* publishes an article by Samuel Epstein that claims that as a result of the use of the drug, milk consumption will increase the risk of breast cancer.

April. Congressional representatives ask the GAO to examine conflicts of interest by FDA employees.

August. Monsanto publishes its analysis of the data from its animal trials in the United States and Europe.

1996 Monsanto completes the post-approval monitoring program (PAMP).

CANADA

1990 Monsanto submits an application for a notice of compliance under the product name Nutrilac.

1993 The Canadian government announces guidelines for the regulation of biotechnology. As with the U.S. Co-ordinated Framework, products will be regulated under existing statutes by the departments charged with administering them.

1994 March. The House of Commons Standing Committee on Agriculture and AgriFood opens hearings on the impact of the introduction of the drug. Concerns are raised about the impact of increased productivity on the supply management system in the dairy industry. The committee requests a legislated moratorium on the drug, and the government comes to a voluntary arrangement with the manufacturers. The government establishes a task force to examine the issues raised by the committee.

November. CBC's current affairs program The Fifth Estate broadcasts a report in which Health Canada scientists allege that Monsanto offered them a bribe in exchange for drug approval. Monsanto officials deny this, stating that they had offered to invest in further research in Canada if the drug was approved.

1995 The voluntary moratorium ends. The National Dairy Council, the National Farmers' Union, and the Council of Canadians request another moratorium. Health Canada requests more animal health data.

1996 Provel, a division of Eli Lilly, asks for its rbGH submission to be placed on hold pending the outcome of Monsanto's submission.

1997 Health Canada scientists claim that their concerns about potential public health risks from animal drugs are being ignored.

1998 The Standing Senate Committee on Agriculture and Forestry begins hearings into the human and animal health aspects of rbST.

1999 14 January. Health Canada announces that it has rejected rbST for licensing in Canada.

Acronyms

CAC	Codex Alimentarius Commission
CDC	Canadian Dairy Commission
BVD	Bureau of Veterinary Drugs (Canada)
bST	bovine somatotropin
CEPA	Canadian Environmental Protection Act
CFIA	Canadian Food Inspection Agency
CMSMC	Canadian Milk Supply Management Committee
CVM	Center for Veterinary Medicine (U.S.)
FAO	United Nations Food and Agriculture Organization
FDA	Food and Drug Administration
FFDCA	Federal Food, Drug and Cosmetic Act (U.S.)
FOI	Freedom of Information
GAO	General Accounting Office
HPB	Health Protection Branch (Canada)
IGF-I	insulin-like growth factor I
INAD	Investigational New Animal Drug Application
JECFA	Joint WHO/FAO Expert Committee on Food Additives
NADA	New Animal Drug Application
NADE	Office of New Animal Drug Evaluations (U.S.)
NBAC	National Biotechnology Advisory Committee (Canada)
NIH	National Institutes of Health
NOC	Notice of Compliance (Canada)
OSTP	Office of Science and Technology Policy
PAMP	Post-Approval Monitoring Program
rbGH	recombinant bovine growth hormone (also known as rbST)

rbST	recombinant bovine somatotropin
TFPC	Toronto Food Policy Council
VMAC	Veterinary Medicine Advisory Committee (U.S.)
WHO	World Health Organization

Science and Social Context

1 Overview

INTRODUCTION

On 14 January 1999 Health Canada rejected Monsanto's application to license recombinant bovine growth hormone (rbGH) in Canada. (The drug is also known as recombinant bovine somatotropin (rbST) or simply as bovine somatotropin (bST)).[1] The product is developed using recombinant DNA (rDNA) technology, a form of biotechnology. When injected into lactating dairy cows, the product results in a 10 to 15 percent increase in milk production.[2] In the early 1980s four multinational chemical and pharmaceutical companies competed to bring the product to market – Monsanto, a chemical corporation turned biotechnology company based in St Louis, Missouri; Elanco, the animal products division of Eli Lilly, a pharmaceutical corporation; Upjohn, also a pharmaceuticals producer; and American Cyanamid, a pesticides producer. By 1998, however, only Monsanto's application was active in Canada.

Health Canada's decision, which was announced after nine years of review, was made on the grounds that although the drug posed little risk to human health, animal health was jeopardized. The last six months of the review were particularly turbulent, as Senate hearings probed the decision-making process and scientists alleged that they had been pressured by the company and senior management to approve the drug in spite of their own misgivings.

Health Canada's rejection of the drug stands in contrast to the approval by the U.S. Food and Drug Administration (FDA) in 1993. The difference between the American and the Canadian regulatory response is one of the puzzles that this book explores. However, this puzzle is intended to illustrate not only differences between Canada and the United States, but also the nature of the relationship between science and policy. It is the production and interpretation of scientific evidence in the policy context that is the focus of the book.

I argue that although there was agreement among the various actors regarding what the evidence *showed*, the *interpretations* of that evidence diverged depending on the context from which it was viewed. That is, the policy implications of the scientific evidence differed depending on whether the scientists involved had regulatory responsibilities and, if so, on which country they were located in. In making this argument, I rely on Helen Longino's (1990) concept of contextual empiricism. Longino suggests that judgments about empirical evidence are based on background assumptions we bring to its assessment; and these assumptions are related, in turn, to the context in which those judgments are formed. In *Science as Social Knowledge* Longino argues that scientific practice involves two types of values, constitutive values and contextual values. Constitutive values are the goals that science seeks to attain, goals such as truth, scope, accuracy, and fruitfulness. They are "the source of the rules determining what constitutes acceptable scientific practice or scientific method." Since these goals are assumed to be the prerogative of science, they are often not regarded as values. Contextual values, on the other hand, "belong to the social and cultural environment in which science is done"; they consist of "group or individual preferences about what ought to be" (4).

Longino argues that there is a necessary connection between the background assumptions that we bring to the reading of scientific – and everyday – evidence and the conclusions we draw from it. Consequently, she states that "evidential reasoning is always context dependent, that data are evidence for a hypothesis only in the light of background assumptions that assert a connection between the sorts of thing or event that the data are and the processes or states of affairs described by the hypothesis" (215). Background assumptions do not always encode social values, but they do provide a means by which these values may enter the reasoning process. Although what counts as "evidence" for a particular hypothesis will depend on background assumptions, these assumptions can still be differentiated from the evidence and the hypothesis that it has been taken to support.

I use Longino's concept of contextual empiricism to analyze the rbGH case. The interpretation of the rbGH evidence depended on two kinds of assumptions: those based on what Kuhn (1970) terms "normal" science – that is, on conventional scientific knowledge as expressed in the literature and communicated in scientists' training – and those based on judgments about the context into which the product was to be introduced. I use the term contextual knowledge, rather than contextual values, in order to capture the nature of the latter kind of judgments. These judgments depended on *knowledge* about practices within that context, as well as *implicit* values about their acceptability. In order to assess the likely effect of a product, scientists must have some knowledge of the social context into which the drug is to be introduced and an implicit acceptance of the values inherent in that context. Therefore, I argue that in order to understand the relationship between science and policy, we must look at the expectations and assumptions that guide the interpretations made by scientists. The decisions made in the rbGH case were ultimately determined by scientists' interpretations. In the United States, FDA scientists concluded that the use of the drug did not represent a risk to human health and that the risk to animal health was minor and could be managed by farmers; the drug was therefore approved. In Canada, on the other hand, Health Canada decision makers agreed with their FDA counterparts on human health but did not agree that the animal health risk was manageable; the drug was therefore rejected after the conclusions on both human and animal health were endorsed by two external scientific panels.

Given this focus on scientists' perceptions, it is possible to argue that the outcome could be explained in terms of the individual personalities involved. However, I argue that these perceptions are shaped by context in two different senses. First, the goals and mandate of the institution in which the scientist operates exert a particular kind of pressure on him or her, thereby promoting or constraining particular types of choices. Second, the broader political-economic context also affects scientists' interpretations. The two factors that were most significant here were the nature of the dairy system and the significance of the biotechnology industry.

With regard to the first sense in which context shapes interpretation, regulatory institutions and processes exert a particular kind of pressure on scientists. Scientists in all contexts – regulatory, corporate, and academic – start from a base of existing scientific knowledge, or "normal" science. The extent to which scientists question the assumptions of normal science is affected by the context in which they operate, however. The mandate of regulatory scientists precludes basic research, and their communication with individuals outside the regulatory body is restricted by the need to maintain the confidentiality of proprietary information. Furthermore, they are conscious of the time limitations that they work within and of the importance of timeliness to their career prospects. They are also conscious of corporate costs and competitive pressures. Regulatory scientists are expected to fulfill their mandate to protect public health, on the one hand, and to avoid imposing an undue regulatory burden on corporations, on the other. They are therefore caught in a juncture between the requirements of the regulatory body and broader corporate and public pressures.

Under these circumstances, regulatory scientists decide what kind of evidence can "reasonably" be requested from the company. That is, their definition of safety is bounded by distinctions between what is reasonable and unreasonable, which in turn reflect distinctions made in conventional science. If the risk can already be explained on the basis of existing knowledge, they are unlikely to ask for further data. In the rbGH case, existing scientific knowledge about the category of hormones to which rbGH belonged – protein hormones – and about natural growth hormone indicated that it was unlikely that the product would represent a risk to human health. In both Canada and the United States, therefore, it was decided that long-term human health testing was unnecessary. Although an additional two-week feeding study was conducted, it was done as a result of political pressure rather than scientific doubt.

Generally, the resulting data confirmed scientists' expectations, although anomalous results were reported. But the anomalies were apparent only in contrast to an expected pattern that had already been established by earlier studies. In instances where such anomalies did appear, they were also viewed through the conventional paradigm, in order to decide what effects could reliably be attributed to the drug itself. Regulators distinguished between *statistical* and *biological* significance. That is, they did not assume that a statistically significant difference between the treated and control groups could automatically be attributed the drug but considered whether the changes were consistent and whether they made sense in terms of existing scientific knowledge about its potential effects.

This decision not to proceed with long-term human health testing was questioned by critics and in Canada, in particular, by scientists within the Health Protection Branch (HPB). Although scientists within the HPB agreed that it was unlikely that rbGH would result in human health problems, this conclusion had not been demonstrated empirically with long-term testing. They argued that the anomalies that had appeared in short-term tests, moreover, warranted further examination. Because rbGH is a nontherapeutic drug with no benefits to consumers and because milk is consumed over a lifetime, scientists within the HPB who reexamined the department's decision argued that it needed to be investigated far more thoroughly. They also pointed out that recent research challenged earlier assumptions about the potential impact of the drug. On the other hand, reviewers responsible for the decision and the external panels that reviewed their work argued that since most of the anomalous results were seen at dosages much higher than those to which humans would be exposed, further testing was not warranted and would create a higher standard for the regulation of rbGH than for other food products. Although recent research raised questions about earlier assumptions, it did not challenge them sufficiently to justify banning rbGH. The external panels argued that because absolute safety could never be achieved, regulatory resources should not be directed to cases where the risk is low. Scientists criticizing the decision, however, pointed out that long-term effects were unknown and that given the lack of benefit to consumers, such a risk should not be undertaken. In both instances, the scientists' positions depended on their view of the role and responsibility of regulation and the regulatory body.

The approach taken by regulatory scientists can be contrasted with that of scientists engaged in basic research, who are more likely to question the assumptions from conventional science and to require empirical support for them. Basic scientists are also more likely to acknowledge the limitations of their experimental method, rather than to make comparisons with the existing context in order to generalize from limited experimental data. Corporate scientists, on the other hand, are also likely to rely on conventional knowledge and to rely on contextual knowledge to downplay risk and ambiguity even further. For example, in the rbGH case, corporate scientists claimed not only that was there no risk from increased levels of growth factor in milk but also that there was in fact no increase in growth factor in milk. In this study, I examine the methodologies and conclusions of corporate and academic science in order to contrast them with the regulatory review process and in order to explore the development of positions that fed into the policy debate. I also examine the interpretation of academic scientists who were responsible for conducting drug trials under contract to the product manufacturers.

Unlike basic scientists, regulatory scientists are restricted in their ability to communicate with scientists outside the regulatory agency and the corporation. The confidentiality of data submitted for regulatory review creates problems both in the relationship between regulators and the rest of the scientific community and in the relationship between regulators and the public. In the case of rbGH basic scientists' questions about the FDA's conclusions could not be answered with the limited information that was released, and the public's concerns were aggravated when the release of information conflicted with the story presented by other actors in the debate, particularly academic scientists and health professionals. University extension workers had promoted the product as "safe" before the FDA had completed its review, but leaks of data and subsequent official releases of information that appeared to contradict this finding eroded confidence in both the regulator and other public bodies. The rbGH case shows that the complexity of scientific information is likely to be distorted in the political debate and that the communication of risk during the review process is inherently problematic under the current system.

The relevance of the second factor that affects scientists' interpretation of the evidence – the broader political-economic context – can be seen in the decision on animal health, where the difference between the United States and Canada becomes salient. Reviewers in both countries decided that the animal health data showed problems that were both statistically and biologically significant. The scientists differed in their assessment of the extent of the problem, however. In the United States, animal health effects were also determined to be manageable, a judgment that depended on knowledge about existing dairy practices and the implicit acceptance of their viability. All technologies require us to adapt our behaviour in some way. In the case of rbGH, it reinforced a system of dairy production that relies on technologies such as antibiotics and reproductive drugs. Existing dairy practices were factored into the studies of rbGH, in which reviewers required a balance between a "scientific" study and a "realistic" study that would reflect actual conditions on the farm. Those actual conditions were taken to include the use of what are known as "extra-label" drugs, that is, drugs that have been used in ways that violate their conditions of approval. The data from these studies still showed that problems increased, but it was concluded that a "successful" farmer who was managing existing problems could also manage these problems as well. A successful farmer, as a series of Office of Technology Assessment (OTA) reports pointed out, was located in the south or southwest of the country and operated an industrial dairy of between five hundred and fifteen hundred cows. When you have more animals in larger units such as these, the possibility of disease is higher, and so is the resultant use of antibiotics. Since the rbGH decision, the United States has also introduced legislation that has legitimized the use of extra-label drugs. The FDA's assessment placed the ultimate responsibility for managing animal health problems with the farmer and with the institutions monitoring drug residues in the milk supply.

In Canada, on the other hand, animal health problems were regarded as "severe," rather than subtle, and it was the regulators, rather than the farmers, who were seen as responsible for protecting animal welfare and the viability of Canadian farms. The consequences of increased disease rates were viewed in terms of the supply management system; in Canada, if a milk tanker is found to contain an unacceptable level of antibiotic residues, the entire tanker of milk is disposed of and the farmer is responsible for the cost of the milk – up to $18,000. If, in order to avoid tanker contamination, the farmer disposes of his or her milk before pick-up, he or she may not be able to meet quota requirements.

The animal health decision and the reaction against rbGH need to be placed in the context of the farm crisis of the 1980s. The literature on the political economy of agriculture (see, for example, Friedmann 1991, 1994; McMichael 1992; Goodman 1987) has explained the crisis as the result of serious problems that emerged with the industrial model of production into which agriculture has been integrated. In the 1930s national governments introduced farm support programs to protect farmers from the vagaries of the international marketplace and to reduce the disparity between urban and rural incomes. The United States introduced price-support programs, and other countries adopted similar programs modelled on the American system. Subsequently, the post–World War II economic order encouraged the development of policies for the national regulation of agriculture. This model, paradoxically, encouraged both the surplus production of food and the growth of transnational agribusiness firms who supplied the industrial inputs for farm production and purchased the resulting produce in the form of inputs for processed food. In the early 1970s, however, this system collapsed with the export of massive quantities of wheat to the Soviet Union (the socialist bloc had previously been excluded from the export of surplus food), the breakdown of the Bretton Woods monetary and trade regimes, and the oil price hike. Food surpluses returned in the 1980s, however, and the glut on the global market led to decreased prices and lower profitability for farmers. Governments began examining alternative modes of regulation: for the first time since negotiations began, agriculture was included in the Uruguay Round of negotiations for the General Agreement on Tariffs and Trade (GATT), with the United States aiming to reduce agricultural protection schemes worldwide.

In this context, representatives from agribusiness corporations, including the proponents of rbGH, argued that rather than protecting agriculture, governments needed to facilitate competitiveness in a globalized marketplace. Agricultural biotechnology could promote the productivity and competitiveness of American crops in world markets, a theme that was reiterated repeatedly by rbGH proponents at successive congressional and advisory committee hearings in the United States. With global competitiveness as the goal of the early 1980s, agricultural economists suggested that rather than banning the technology in order to preserve the price-support system in the United States, or the supply management system in Canada, these programs should be eliminated in order to facilitate the application of the technology and the increased productivity of North American agriculture. The solution to the overproduction crises that resurfaced in the 1980s was not to block productivity-enhancing technology but to eliminate programs that distorted market mechanisms. Most rbGH proponents accepted, however reluctantly, that the level of government support for agriculture would gradually decline and that the trend toward greater efficiency, fewer cows, and fewer farms would continue. Under these conditions, rbGH represented a logical means to increase the productive efficiency of American agriculture and its capacity to compete in world markets. The manufacturers of the drug viewed technological applications as just as essential to the continued economic competitiveness of agriculture as they were to other industries. Corporate developers believed that the decline of American industry could be attributed to its failure to invest in new technologies and that agriculture was similarly doomed if it resisted the advent of biotechnology.

Farm and rural advocacy groups resisted this argument, however. The potential impact of rbGH on the agricultural system, as well as on health, was debated in several fora made available through the legislative branch. In both countries concerned farm, consumer, and food-policy groups and antibiotechnology activists used the available political channels to investigate the regulatory body's evaluation process, request third-party reviews, and impose voluntary or legislated moratoria. Due to the nature of the political system in the two countries, there were more avenues available to critics in the United States than in Canada (see chapter 3 and Jasanoff 1986, 1990, 1995b). The fragmentation of political power in the United States means that the regulatory process is closely supervised by both Congress and the courts. Congressional committees may call public hearings on particular issues, and congressional representatives may request investigations by the General Accounting Office (GAO) or by the inspector-general of the relevant department. Once made, the regulatory decision may be reviewed by the judicial branch.

In the rbGH case, the oversight by the legislative branch in the United States did not change the outcome of the case. It did, however, force regulators to defend their decision before the public, which may have slowed the pace of review and may arguably have made reviewers more determined to defend their decision. Where intervention by the legislative branch did have the greatest impact was on the question of the indirect human health impact of increased animal illness. The GAO argued that this question had not been adequately addressed by the FDA, which then examined the issue and had it reviewed by its Veterinary Medical Advisory Committee (VMAC). However, the VMAC endorsed the FDA's decision that the risk was "manageable."

In the Canadian case, there were fewer channels open to critics of the process; however, due to the controversial nature of the product, the subject was investigated by a House of Commons committee and, after Health Canada scientists had raised their concerns about the review process with the media, by the Senate. Ultimately, however, the decision reflected the initial reviewers' concerns as confirmed by two external panels.

The use of rbGH was controversial internationally as well as domestically. Like Canada, the European Union rejected the product on animal health grounds in 1999. A committee struck in the late 1990s also raised human health issues similar to those discussed by Health Canada scientists. In the early 1990s the EU had accepted the safety of the product but had rejected its sale on socioeconomic grounds. By the end of the decade, however, a different scientific review structure had been established in Europe, and new law made it possible to reject nontherapeutic drugs on animal welfare grounds. Although the EU had not banned imports of dairy products from the United States on the grounds that they might contain rbGH, throughout the 1990s it had voted against the establishment of a standard for the drug at the international standard-setting body, the Codex Alimentarius Commission. Such a standard would not have forced the EU to license the sale of the hormone. It would, however, have prevented it from banning the import of U.S. dairy products on health and safety grounds, and it may have increased the pressure on governments to license the drug. The United States had favoured the adoption of a Codex standard; but at the Codex meeting in 1999 it accepted that there was no consensus on the issue and did not pursue it further.

Sheila Jasanoff has said that "regulation ... is a kind of social contract that specifies the terms under which state and society agree to accept the costs, risks and benefits of a given technological enterprise" (1995a, 311). This book explores the problematic nature of this contract and the difficulty for regulators of assessing the risks associated with a technology when there is no consensus about its benefits.

METHODOLOGY

In order to explore the social and political background to the decision-making process regarding rbGH, as well as the process itself, I examined U.S. and Canadian government documents pertaining to both the facilitation and the regulation of biotechnology, including relevant statutes and administrative guidelines, and, in the U.S. case, congressional hearings regarding particularly controversial decisions and trends in biotechnology, such as the patenting of animals and the commercialization of scientific research. In regard to the rbGH debate, I also researched the congressional and parliamentary hearings on the subject and, for the United States, the various advisory committee meetings regarding human health implications, labelling issues, and the post-approval monitoring program. I also covered the various inspector-general's reports of the General Accounting Office and the Department of Health and Human Services and, for Canada, reports of the House of Commons Agriculture and AgriFood Committee, the government response, and the rbST task force reports.

The political economy of biotechnology development was researched through an examination of documents from the Monsanto company, including annual reports and internal publications; newspaper articles, particularly from the *Wall Street Journal* and the chemical trade press; and the secondary literature.

I also obtained reports, newsletters, and copies of correspondence from the various environmental, farm, and food-policy groups opposed to the drug, including the Canadian Institute for Environmental Law and Policy, the Toronto Food Policy Council, the Ram's Horn, Rural Vermont, and the U.S. National Farmers' Union.

Information regarding the scientific debate was obtained primarily from the FDA's Freedom of Information Summary and an earlier summary of the human health data and its interpretation, published in *Science* in August 1990. I also read articles from scientific journals, including *Science*, *The Lancet*, *Physiological Reviews*, the *Journal of Dairy Science*, and the *Journal of the American Veterinary Medical Association*. And I attended the final meeting of the FDA's Veterinary Medicine Advisory Committee (VMAC), which reviewed the data from the rbGH post-approval monitoring program in November 1996.

My conclusions are based primarily on data from interviews conducted between August 1996 and August 1997. I interviewed twenty-six individuals, several of whom were interviewed twice. While most of the interviews were conducted face-to-face, some were conducted over the phone, due to geographical or time constraints. I interviewed two scientists from the FDA who were prominent in the rbGH evaluation and two other FDA officials. My informant regarding the Canadian regulatory process was formerly with the Bureau of Veterinary Drugs. I spoke with four Monsanto scientists and a Monsanto official. I also spoke to scientists at universities in Canada and the United States who acted as principal investigators for the safety and efficacy trials, one university scientist who had been critical of the process, and members of professional associations (such as the American Medical Association) who authored position statements endorsing the product. In addition, in order to contrast their methodology with that of scientists in the rbGH debate, I interviewed biomedical researchers outside the debate who were investigating growth factor physiology. The book does not include the names of the scientists and officials interviewed. However, this information is included in the dissertation on which the book is based (see Mills 1999).

I had no difficulty obtaining interviews with scientists outside the debate. But since the drug was still under review in Canada during the period of my research and since the confidentiality of the information submitted in support of the drug application was therefore maintained, I was unable to obtain interviews with current Health Canada scientists. I was also unable to obtain interviews with some critics of the drug and with one of the university scientists whose perspective I particularly wanted to hear. The FDA and Monsanto were initially reluctant to participate. They received copies of my questions before participating, and the FDA's counsel reviewed them beforehand. Interviewees were instructed that they could refuse to answer any question. They rarely took up this offer but often were unable to provide specific details in response to a particular query, either because answering would have risked releasing proprietary information or, given the amount of time that had elapsed since the initial decisions were made, because the details were beyond recall. Quotes attributed to particular individuals were sent back to them for confirmation and correction, and most of the interviewees replied with minor alterations.

A standard format provided the basis for the interviews. The questions focused on whether a literature review had been conducted and how it had been decided what literature was relevant; how the scientists had decided what kinds of studies should be conducted; how they had proceeded with the analysis of the data; what time period they had worked within, and how this had affected their analysis; what kinds of evidence would have enabled them to conclude that a product was "safe" or "unsafe"; what kinds of interaction they had had with other organizations or groups; and whether this had influenced their decision-making. This format was heavily influenced by the first interview, with a former FDA official, who raised many of these issues in the course of the discussion.

In addition to these basic questions, I asked very specific questions regarding aspects of the research particular to the institution. For example, I asked FDA scientists about the data and conclusions reported in the article in *Science* in 1990, and the Freedom of Information Summary of 1993; Monsanto scientists were asked about their published articles and analysis of the data submitted to the FDA, as were university researchers; biomedical scientists were asked about the content of their own research. I also asked Monsanto scientists and officials about the process of drug development, their reaction to the public controversy, and their interaction with professional associations on this issue.

As well as interviewing scientists and regulators, I spoke to members of Rural Vermont, a farm group opposed to the introduction of the drug, and to a member of the Vermont legislature. These interviews were not structured and related primarily to Rural Vermont's interactions with the FDA, the reasons for its opposition to the drug, and the actions it had taken since it was introduced.

Communicating the scientific debate in a way that was understandable and yet that did justice to the complexities of the debate was one of the challenges of this book. Researching the book was also challenging because the controversial nature of the subject and the confidential nature of the data made it particularly difficult to obtain interviews and to obtain complete information once interviews were given. Interviewees often could not refer to specific details of the case but spoke in generalities. The content and detail of discussions within and between organizations could not be divulged. The focus on rbGH alone, as in a focus on any single case study, places certain limitations on the generalizability of the conclusions that can be drawn here. On the other hand, the very uniqueness of the case and the confidentiality of specific details offer greater scope for generalization because interviewees spoke generally rather than specifically.

The first three chapters of this book outline the different contexts in which rbGH science took place. Chapter 2 locates the development of rbGH within the broader political economy of agricultural biotechnology. It outlines Monsanto's transformation from a chemical producer to a biotechnology or "life sciences" company and locates this transformation within the broader restructuring of the chemical, oil, and pharmaceutical industries, which have invested in biotechnology as a strategy to compensate for declining profits since the oil crisis of the 1970s. It explores the company's rationale for pursuing the development of rbGH, its relationship with the agricultural colleges and land-grant universities in the United States and Canada, and its reliance on third-party scientific groups and university scientists to endorse and promote the safety of the drug. Chapter 3 outlines the U.S. regulatory context, including the biotechnology policy framework, the regulatory requirements for the evaluation of animal drugs, the use of legislative and judicial mechanisms by those opposed to the drug, and the history of the controversy. Before the introduction of the drug, the Reagan administration had introduced several measures to promote technological innovation by the private sector and by the biotechnology industry, in particular. It also sought to reduce regulatory restrictions on the industry. Consequently, there were few avenues for debating biotechnology in general; instead, the health and safety implications of particular drugs such as rbGH were debated. Although there were more avenues for debating and protesting against the introduction of rbGH in the United States than in Canada, the political structure did not have an impact on the FDA's decision. Chapter 4 outlines the context of the decision-making process and the controversy in Canada. Although the Canadian government has also sought to promote the biotechnology industry, it has not gone as far as its counterpart in the United States. The rbGH debate in Canada was also concerned with the implications of the drug for the country's supply-management dairy system, and scientists' interpretation of the implications of the data for farm and animal welfare proved decisive. In the rbGH case, fora within the legislative branch were used to examine the impact of the drug, and they draw attention to scientists' concerns about the human health implications of this drug, as well as of other animal drugs approved against their advice. However, the final decision reflected the original assessment by Health Canada scientists. Chapter 5 analyses the scientific debate around both the human and animal safety of rbGH, outlining the position taken by each institution – primarily the FDA and Health Canada, as well as the universities, Monsanto, and the critics – and explaining the reasoning behind each position. Although there was a consensus about the findings from the safety and efficacy trials, the weight given to certain aspects of the findings – and the conclusions about their implications – differed. Chapter 6 analyzes the findings from the chapters presenting the data and concludes that difficulties in the science policy relationship arise at the point where judgments about the data are made. Although there may be widespread agreement about the meaning of the data, its interpretation and social meaning depend on the context from which it is viewed.

2 The Economic Context: The Political Economy of Agricultural Biotechnology

INTRODUCTION

The purposes of this chapter are threefold. First, it sets out the economic context by providing an overview of the development of the biotechnology industry in the United States, and its facilitation by government policies. The success of the biotechnology industry was regarded as crucial to the United States' economic future. Although government policy fostered the creation of the industry in the 1990s, agricultural biotechnology came to be dominated by multinational corporations such as Monsanto that were exporting their innovations and investing in seed companies and biotechnology firms located around the globe. However, protests by the environmental movement and subsequent consumer boycotts in Europe have brought an abrupt change in Monsanto's fortunes and the future of agricultural biotechnology is now uncertain.

Second, this chapter provides a brief outline of the debate about the economic impact of rbGH. The application of technology in dairy farming had led to a steady increase in the milk supply, with the result that huge oversupply problems emerged in the 1980s. Government support for surplus production was declining, however. Agricultural economists suggested that the dairy industry would be transformed by the application of biotechnology and that resulting oversupply problems could be eliminated through reliance on market mechanisms rather than government support. However, this conception of adjustment was resisted by many farmers and farm organizations in Canada and the United States.

Third, the chapter outlines the relationship between Monsanto, the university scientists who conducted its animal safety and efficacy trials, and the medical associations that endorsed its product.

The exploration of the economic context provides the background for understanding Monsanto's perspective on the debate about safety, and the social and economic issues that fed into the broader policy debate. Since regulators were sensitive both to the need to protect public health, on the one hand, and to corporate costs and competitive pressures, on the other, an overview of the economic context is also necessary to understand what these costs and pressures were.

REGULATORY CHANGES AND THE COMMERCIALIZATION OF GENETIC RESEARCH

The commercial potential of biotechnology was perceived early in its development, and the corporate sector's investment was further encouraged in the early 1980s by a series of regulatory changes instituted by the Reagan administration. These policies were intended to increase America's capacity for technological innovation by encouraging the commercialization of basic research undertaken at universities and federally funded institutions. Commercialization was also facilitated through tax incentives and judicial decisions that granted proprietary rights over previously unpatentable microorganisms, animals, and plants.

The Patent and Trademark Amendment Acts of 1980 (PL 98-260) and 1984 (PL 98-260) aimed to promote efforts to develop a uniform federal patent policy and to commercialize government-funded research by allowing recipients of federal research funds – including universities and small businesses – to patent their innovations (OTA 1990, 55).[1] According to Slaughter and Rhoades, these laws "blurred the boundaries between public and private sectors" and "gave new and concrete meaning to the phrase 'commodification of knowledge.' The act[s] enabled universities to enter the marketplace and to profit directly when universities held equity positions in companies built around the intellectual property of their faculty as well as to profit indirectly when universities licensed intellectual property to private sector firms" (1996, 318).

Cooperation between firms and between government-funded institutions and the private sector was furthered by new legislation. The National Co-operative Research Act of 1984 (PL 98-462) relaxed antitrust law to permit research collaboration among previously competitive antitrust firms.[2] The Technology Transfer Act of 1986 (PL 99-502) allowed government-operated laboratories to enter into cooperative research arrangements with the private sector.Tax incentives also encouraged private R&D investment. The Economic Recovery Tax Act (PL 97-34) granted companies a 25 percent credit for increases in R&D expenditure above base-year expense levels. Relaxation of regulatory guidelines also provided incentives for American companies to manufacture pharmaceuticals and food additives at home rather than overseas. In 1986, the Drug Export Amendments Act permitted drugs that had not been approved for use in the United States to be exported to twenty-one countries (Slaughter and Rhoades 1996, 320).

As a result of policy decisions, the ability to secure proprietary rights over innovations was extended to public institutions, and as a result of judicial decisions, the scope of innovations that could be patented was extended to life forms. Before 1980, living organisms were regarded not as patentable subject matter but as products of nature. However, in 1980 the Supreme Court ruled in *Diamond v. Chakrabarty* that a living, man-made organism (a bacterium) was patentable subject matter within the meaning of the Patent Act (OTA 1987, 7), but the Court did not address the issue of whether plants were patentable subject matter under the act. Previously, plant breeders had claimed intellectual property rights under the Plant Patent Act (PPA) of 1930 or the Plant Variety Protection Act (PVPA) of 1970. General patent law, however, offered broader protection than that available under either of the plant patent acts, and in 1985 the Patent and Trademark Office's Board of Appeals and Interferences ruled in *Ex parte Hibberd* that a transgenic corn plant was patentable subject matter. In 1988, the first animal patent was issued to Harvard University for a mouse with a cancer-causing, or "onco," gene inserted into its DNA. An exclusive license to apply the technology was granted to DuPont, a major sponsor of the Harvard research (OTA 1990, 12).

The negotiation of intellectual property agreements as part of the Uruguay Round of the General Agreement on Tariffs and Trade (GATT) and the North American Free Trade Agreement (NAFTA) ensured that intellectual property claims would be respected outside the United States. Section 5, Article 27 (5) of the international Agreement on Trade-Related Aspects of Intellectual Property Rights (TRIPS) specifies that "Each Party shall make patents available for any inventions, whether products or processes, in all fields of technology, provided that such inventions are new, result from an inventive step, and are capable of industrial application."

However, countries do not have to extend intellectual property protection to plants and animals under the TRIPS and NAFTA agreements. Article 27 (3) of TRIPS allows signatories to exclude the following inventions from patentability: "diagnostic, therapeutic and surgical methods for the treatment of humans and animals" and "plants and animals other than microrganisms, and essentially biological processes for the production of plants or animals other than non-biological and microbiological processes for such production." However, plant varieties must be covered by an intellectual property system either through patents or a scheme of sui generis protection (i.e., special legislation dealing solely with plant varieties) (McMahon 1995, 24).

Signatories to NAFTA may also exclude plants and animals from patentability. Under article 1709(2) of NAFTA, "A party may exclude from patentability inventions if preventing in its country the commercial exploitation of the inventions is necessary to protect *ordre public* or morality, including to protect human, animal or plant life or health or to avoid serious prejudice to nature or the environment for reasons including the protection of human, animal or plant life, provided that the exclusion is not based solely on the grounds that the Party prohibits commercial exploitation in its territory of the subject matter of the patent."

COMMERCIALIZATION AND BIOTECHNOLOGY

The commercialization of biotechnology has been fostered not only by government policy but by the belief that the application of biotechnology would have a revolutionary economic impact. The biotechnology industry consisted initially of small start-up companies funded by venture capital. However, as venture capital sources diminished, biotech start-ups entered into various arrangements with multinational chemical and pharmaceutical giants, including the biotechnology industry leader, Monsanto. Multinationals pursued investments in agriculture, pharmaceuticals, and nutrition – areas known as "life sciences" – on the grounds that recombinant techniques could be applied in all these fields and that the synergies among them would lower costs. However, recent consumer backlash against biotechnology in Europe has undermined the commercial potential of agricultural biotechnology and hence of the "life sciences" concept. Multinationals are currently divesting their agricultural companies and refocusing on drug development.

Investment in biotechnology was initiated and sustained by the belief that it represented "the century's third great technological revolution – after atomic fission and computers" (Naj 1989, B1). In 1984 Monsanto's CEO, Richard Mahoney, said that "Our initial investigation into genetic engineering made it clear it was going to be at least as important to the [chemical] industry as the petrochemical revolution of the 1930s" ("Monsanto's New Regime" 1984, 64).

Biotechnology has also been perceived as another form of information technology, rather than as a distinct technological wave. *Barron's* has noted that "Because genes are snippets of information that can be rewritten to produce various outcomes – starchier potatoes, say, or pesticide-resistant soybeans – agricultural biotech is analogous to computer technology" ("The Ultimate Synthesizer?" 1995, 10). Similarly, Monsanto's CEO, Robert Shapiro, has observed that "Biotechnology is, in a sense, a subset of information technology. Just as information technology is the science of encoding data onto silicon wafers, biotechnology is the science of encoding information into living systems" (Shapiro 1996, 8). By 1997 Monsanto had announced its discovery of the biological equivalent of Moore's law, which predicted that the information-processing capacity of silicon chips would double every eighteen to twenty-four months; the company believed that the speed of knowledge generation in the biosciences was proceeding at the same rate and would have equally dramatic economic consequences (Monsanto 1997).

In the United States, two economic institutions drove the development of the biotechnology industry: the small start-up companies mentioned at the beginning of this section, which were founded by university scientists and financed by venture capital, and multinational pharmaceutical and chemical companies (Kenney 1986, 132). The establishment of small start-up companies devoted to the development of biotechnology was made possible by venture capital investments, and venture capital itself became more formalized in the 1970s: large corporations created divisions that specialized in financing innovative companies with the expectation of realizing a 500 to 1000 percent capital gain when the stock was publicly traded (Kenney 1986, 133, 142). Genentech was one such company. Founded in 1976 by Robert Swanson and Herbert Boyer, formerly a professor at the University of California in San Francisco, Genentech was the first company to be completely focused on biotechnology R&D (156). When a stock offering was made in 1980, prices jumped from $35 to $89 within a day, setting a Wall Street record for the fastest per-share price increase (OTA 1991, 4).

By 1983, however, venture capital sources were no longer as willing to extend further financing; biotechnology start-ups were relying more on stock offerings as a means of raising capital and had also developed new mechanisms, such as the R&D limited partnership, to fund production scale-up or clinical trials. Chemical and pharmaceutical multinationals began investing directly in start-up companies and forming other strategies for gaining access to their innovations. The chemical industry turned to biotechnology in an attempt to reverse the decline in profits brought about by a lack of innovation, increased costs as a result of the oil crises of 1973 and 1979, and the cost of cleaning up environmental damage from earlier product cycles (Kenney 1986, 191–3). By the 1970s chemical companies had become commodity producers; to reduce their dependence on low-profit bulk chemicals, they invested in pharmaceuticals and patentable products with a high profit margin.

Start-ups and multinational corporations entered another series of alliances and equity agreements in the early 1990s, when biotech stocks went through another cycle, falling in 1993 and 1994 (Thayer 1996, 13). In need of further capital, biotech start-ups entered into alliances with pharmaceutical companies: 66 alliances valued at $3.21 billion were formed in 1994, and a further 171 agreements had been reached by the end of 1995 (14). Genentech, which had aimed to be a freestanding corporation (Kenney 1986, 161), sold a 60 percent stake to Roche for $2.1 billion in 1990, along with the rights to acquire up to 80 percent of the company by 1999 (Thayer 1996, 15). A biotech company's value on the stock market was in fact determined by its relationship with a large pharmaceutical firm. The manager of one of the most profitable biotech investment funds thus limited his investments to companies that had entered into partnerships with pharmaceuticals: "The biotech industry is fragmented. Drug development has to be centralized so that resources are properly allocated. For that to work, research should be done by the biotechnology firms, and development should be done by the drug industry" (Brammer 1995, 30). According to fund managers, an alliance with a multinational firm indicates the commercial potential of the start-up's technology (see Brammer 1995,Northfield 1996). Conversely, technological innovation by start-ups is also regarded as crucial for the survival of the multinationals. It has been predicted that "before long, practically every multinational will have allied itself with a genomics company" (Yakaubski and Northfield 1996, B3).

Acquisition of start-up firms by multinationals occurred concurrently with global consolidation of the pharmaceutical industry, as companies developed a "related-products strategy," focusing on particular segments of the business. In 1986 Kenney argued that biotechnology had the greatest potential to increase productivity and transform production in the pharmaceutical, chemical, agricultural, and food processing industries, providing "a common technical base on which [these industries] can be united" (218). This prediction would appear to have been fulfilled with the creation of the "life sciences" industry, in which chemical and pharmaceutical companies have applied recombinant technology in the manufacture of both chemicals and drugs, as well as agriculture and food products. In 1996, Ciba-Geigy and Sandoz combined to form the world's largest agribusiness, Novartis, which is comprised of agricultural, pharmaceutical, and nutrition divisions, and has a $2 billion R&D budget. The following year, Novartis acquired Merck's crop-protection business (presumably pesticides) for $910 million (Bagli 1997, C7). Merck, meanwhile, was undergoing its own restructuring, merging its animal health and poultry genetics businesses with those of Rhone-Poulenc to create Merial Animal Health, the largest veterinary drug company (Young 1997, 14).

However, as a result of declining commodity prices in agriculture and increasing consumer and environmental concerns about biotechnology, the convergence between agricultural biotechnology and other sectors has been called into question. Aventis, a company created in 1999 by the merger of Hoechst and Rhône-Poulenc, has spun off its agrochemical business in order to concentrate on pharmaceuticals (Dow Jones 2000, 1); AstraZeneca and Novartis have spun off their combined agrichemicals group, Syngenta (Agence France-Presse 2000, 3); and as part of its take-over by Pharmacia-Upjohn, Monsanto has divested itself of 20 percent of its agricultural business (this development will be explored further in the next section). However, the Rural Advancement Fund International (RAFI), a rural advocacy group that is particularly active on issues affecting developing countries, has argued that the life sciences model has been deferred rather than abandoned, stating that multinationals will "Keep the ag side of their interests at arm's length. In the long run, they won't drop agbiotech altogether because the synergies are just too lucrative to abandon" (RAFI 2000, 3).

A 95-YEAR-OLD STARTUP COMPANY: MONSANTO AND BIOTECHNOLOGY

As the largest investor in biotechnology (Kenney 1986, 212) Monsanto is both an example and a precursor of change in the structure of the chemical and pharmaceutical industries.[3] Like other chemical and pharmaceutical giants, Monsanto recognized the need to extend the market for its existing products and to create new ones and perceived biotechnology as the solution. Although the company ran into difficulties in the early 1990s, it persisted with this strategy and restructured the company to accommodate it. By the late 1990s, it had released several genetically modified seeds onto the market, and was furiously acquiring other seed firms and agricultural biotechnology companies. This strategy, however, left the company heavily indebted, and, as mentioned in the previous section, it was acquired by Pharmacia-Upjohn in 1999.

Monsanto's foray into biotechnology in the 1970s was propelled by a belief in the potential of molecular biology to create a new generation of agricultural products when the market for chemicals waned and by a desire to extend the market potential of its existing products. Losses in 1968 had provoked the biochemist Ernest Jaworski to think about other means of generating income for the company. "After you've invented all the herbicides you need, all the insecticides you need, all the fungicides, what are you going to do to keep growing? I concluded that a time would come when you couldn't solve all problems with chemicals" (Rogers 1996, 4–5). Jaworski pushed Monsanto to develop molecular biology applications in its agricultural division. He had researched how glyphosate – the key ingredient in Monsanto's top-selling herbicide, Roundup – kills plants, and he had predicted that biotechnology would be used to create herbicide-resistant plants. Eventually, Monsanto scientists succeeded in producing crops that were resistant to Roundup, and because farmers were then able to apply the herbicide directly on top of Roundup-resistant crops, allowing for greater weed control without damaging the plant, the market for the herbicide was extended at the same time as a market for the resistant seeds was created. In 2001, sales of the company's glyphosate-based herbicide, Roundup, still accounted for approximately half its revenue (Barboza 2001, C1).

Monsanto's relationship with universities and start-up companies and the creation of its own life science laboratories, parallels the development of the industry. Kenney has identified four patterns of multinational investment in biotechnology. The multinational may, first of all, finance biotechnology research at universities and then obtain an exclusive license to market the technology that the university has developed and patented. Second, the larger corporation may contract with a start-up company for the production of patented inventions that it develops and markets on a commercial scale. Third, it may purchase equity in a start-up company, which keeps it up to date on innovations and rival products. And fourth, it may develop an in-house research capacity, either by establishing its own laboratory or by buying a start-up (1986, 199). Monsanto has engaged in all four patterns of multinational investment.

In the early and mid-1970s Monsanto's pharmaceutical development was pursued through its university agreements. In 1974 the company's vice-president of technology, Monte Throdahl, finalized an agreement with Harvard Medical School that became a harbinger of future university-corporate research agreements (Rogers 1996; Kenney 1986, 60). Under this arrangement, Monsanto provided two Harvard researchers with $21 million over twelve years. As a result, Harvard changed its policy to permit licensing of patents for profit; previously, patents for therapeutic products could only be dedicated to the public. Incidentally, unlike National Institutes of Health grants, Monsanto funding was not dependent on peer review (Kenney 1986, 59)).

Subsequently, Monsanto looked to other research institutions with whom it could reach a more beneficial arrangement than it had reached with Harvard, and in 1981 it contracted with Washington University in St Louis: "Determined to make alliances that would be more useful to Monsanto, [Howard] Schneiderman [senior vice-president of research] took care to structure the Washington University relationships so that both parties would profit; it called for Monsanto to invest $23.5 million over five years to establish a program at the Medical School to discover, study, and isolate proteins and peptides regulating cellular functions." In return, Monsanto received exclusive rights to license any inventions patented by Washington University, and Monsanto scientists and technicians learned new techniques and collected information from the university (Kenney 1986, 67). The terms of this agreement were more favourable to Monsanto's interests than the Harvard agreement, because "The Harvard researchers had been unwilling to share any of the results of their research with Monsanto personnel before [they were] available through open publication. The informally worded Harvard agreement, intended to provide a window into new technology, turned out to be a closed door to Monsanto" (Leonard-Barton and Pisano 1990, 5).

Schneiderman's successor as senior vice-president of research, Dr Philip Needleman, subsequently emphasized the role of universities in Monsanto's research strategy:

> The most important external vehicle for Monsanto's discovery base is the university affiliations they have developed. The university professors have no development costs. All their government grants are pure discovery money. So bang for buck, you have access to some of the finest minds. Monsanto's investment in the university research provides a scientific base, a lead for new discoveries far in excess of the cost of investment. There is no small time player if you are going into biotechnology. Either you can clone the genes, have mammalian cell culture, can build vectors, do all the sequencing – or you are a bit player. (Quoted in Leonard-Barton and Pisano, 1990, 12–13)

In the late 1970s Monsanto began pursuing other strategies for biotechnology development: negotiating contracts with start-up companies as well as universities, contracting with Genentech to produce rbGH in 1979, and entering into further contracts in the early 1980s. It also pursued equity agreements, purchasing a 12.5 percent stake in Biogen and a 30 percent share in Collagen in 1980 (Kenney 1986, 213).

Monsanto also began developing its own in-house research capacity. Howard Schneiderman had been hired in 1979 to direct all corporate research (he was a former dean of the School of Biological Sciences at the University of California at Irvine) (Rogers 1996, 7), and the company opened its own molecular biology laboratories in 1981. Schneiderman cultivated members of upper-level management and gained the support of Richard Mahoney, who became CEO in 1984. The company's in-house capacity was expanded further in that year with the opening of a $150-million life sciences centre (16) and with the acquisition of the pharmaceutical company, G.D. Searle, which was purchased for $2.8 billion in 1985 (Moody's 1995, 4661). As CEO, Mahoney continued to foster the company's biotechnology strategy, which was based around three expectations: that Roundup herbicide would remain profitable throughout the 1990s; that agricultural biotechnology would revolutionize agriculture; and that the company's investment in Searle, which exceeded the division's revenues, would eventually be returned as a result of new product development (Shapiro 1996, 3).

The wisdom of Monsanto's shift in direction was questioned in the early 1990s. In 1991, *Business Week* wondered whether the company was "burning money" with its biotech investments. Monsanto had not yet had a biotech product approved for commercial release, and in 1990 its earnings dropped 20 percent (Siler and Carey 1991, 74). One year later, the company eliminated 10 percent of its workforce – 3,200 people – and wrote down a portion of its inventories of rbGH, Nutrasweet, and pharmaceuticals (McMurray 1992, A3).[4] These losses took place when the profitability of the entire biotech revolution was being called into question. At least twenty-four biotech companies had filed for bankruptcy protection in 1988, and the value of biotech stocks was declining before products or processes had been commercialized. *The Wall Street Journal* suggested that the failure of biotech companies was brought about by a set of conditions that were different from the conditions that had been responsible for commercial failures in earlier technological revolutions: "In the waves of technology that spawned the steel, auto and electronics industries, business failures usually had to do with competition for markets. But biotechnology, which promises to use living organisms to deliver new drugs, crops, fertilizers and pesticides, has run afoul of public fears, regulation and patents confusion" (Naj 1989, B1).

Monsanto continued its biotechnology investment, however. By 1996 the company had acquired three biotechnology start-up companies and another pharmaceutical company, entered into research partnerships with several biotechnology and pharmaceutical companies, and made equity investments in a further three companies, including a 54.6 percent stake in Calgene, producer of the genetically modified Flavr Savr tomato (Monsanto 1997a, inside cover; Miller and King 1995, B16). In February 1996 it invested $160 million in Dekalb Genetics to collaborate on agricultural biotechnology research, including research into the genetic alteration of corn and soybean seeds. According to William Young of the investment firm Donaldson, Lufkin and Jenrette, Monsanto "is almost a cult stock now." George S. Dalman, an analyst at Piper Jaffray, observes that "Monsanto is probably one of the best ways to invest. They understand that technology is a tool that can produce profits by improving existing products, not just creating new ones" (Wyatt 1996, F11).

In 1998 Bill Freiberg, publisher of the Agbiotech Reporter, pointed out that "all meaningful biotechnology research is being done by these 8 to 10 giant agricultural chemical companies in conjunction with seed companies ... I don't know of any of those early start-up companies that even made one dime of profit on their own." There are now few independent agricultural biotechnology companies left; most have been acquired by multinational firms or have been unable to survive without majority equity ownership by a multinational. In 1998 a San Francisco merchant bank did create a venture capital fund for agricultural biotechnology start-ups to counter the domination of the field by a small number of firms, particularly Dow Chemical and Monsanto. But even in this case initial injections of capital for the new fund were supplied by European multinationals Bayer and Agrevo (the latter company was created by Hoechst and Schering) (Pollack 1998, C12).

As Monsanto redirected its efforts toward biotechnology and pharmaceuticals, it shed its petrochemicals and plastics commodities businesses and its oil and gas operations (Reish 1995, 13). Restructuring was completed in 1996, when the company announced that it would spin off its chemical businesses and split into two independent, publicly traded companies. Its herbicide business, the company's most profitable one, remained within Monsanto life sciences (Ewing 1996, A4). A life-sciences company, chaired by Bob Shapiro, retained the Monsanto name, and the chemical business was named Solutia (Monsanto

1997b, A14, A15). As a result of the split, between fifteen hundred and twenty-five hundred Monsanto employees lost their jobs (Ewing 1996, A4). The change was explained by a Monsanto vice-president, A. Nicholas Filipello, who argued that "Chemicals and life sciences are entirely separate businesses, requiring different management and attracting different investors" (Deutsch 1998, C3).[5] Another Monsanto representative explained that a biotechnology-based business requires a different management strategy than a chemical company requires because "The aim with commodity-type chemicals is driving the costs down, efficiencies. You have to be much quicker with biotechnology, any area where you're at the forefront. A year behind and you're out of the competition. You have to try to make it a leaner organization, and more responsive, and it's a different mentality. Old products have been around for 30 years. It's more important for people in that area to spend three months putting together a plan to get a competitive edge. In the area I'm in you need to work quickly to get patents on new products."

The new Monsanto life sciences business was organized into "sectors" rather than units: an agricultural sector, a food and consumer sector, a pharmaceutical sector, a health and wellness sector, and a sustainable development sector. Its staff were organized into teams: "core capability" teams were skilled in scientific knowledge, information technology, and management; "foundation" teams were skilled in law, regulatory affairs, finance and administration; and a "global" team was responsible for identifying business opportunities in developing countries (Monsanto 1998a, 6).

Food. Health. Hope. Monsanto's new corporate logo, which was announced in February 1998, summed up not only Monsanto's metamorphosis from chemical producer to biotechnology investor but the metamorphosis of a generation of chemical, pharmaceutical, and agricultural companies that have diversified into traditionally demarcated areas.[6] Monsanto's position in agricultural biotechnology had increased dramatically by mid–1998. In May the company reached an agreement with Cargill, a U.S. agribusiness firm that currently controls the production and processing of millions of acres of crops. Analysts predicted that this agreement would allow Monsanto to market genetically engineered seed through Cargill's distribution networks; the final crop could therefore also be processed by Cargill (Kilman and Warren 1998, B8). By the end of June 1998 Monsanto had acquired Cargill's seed businesses in Central and Latin America, Europe, Asia, and Africa for $1.4 billion. The deal combined Cargill's distribution network and its seed resources with Monsanto's biotechnology capabilities. In addition, Monsanto had acquired a number of seed companies, including

DeKalb Genetics, Delta and Pine Land, and Holden's Foundation Seeds, which led John McMillan, an analyst for Prudential Securities, to dub the company "the Pac-Man of the agricultural industry" (Reuters 1998a, B13). Holden's Foundation Seeds, one of the few large independent seed companies in the United States, was acquired for $1.02 billion early in 1997, thus providing access to Holden's store of corn germplasm and to networks for distributing genetically modified corn (Rotman 1997, 7). De Kalb Genetics and the Cargill seed companies are two of the ten largest seed companies in the world; Delta and Pine Land is the largest cotton seed company.

The rapid consolidation of seed companies is part of a broader trend in the industry; between 1972 and 1988 multinational drug companies acquired shares in sixty seed-producing firms (OECD 1988a, 28). Ten companies now account for 30 percent of the $23-billion global seed trade (RAFI 1998a, 2); Monsanto's purchase of seed holdings placed it as the world's second largest seed company (RAFI 1998a, 2). The Cargill acquisition put Monsanto just ahead of DuPont in the race for dominance of agricultural biotechnology, providing Monsanto with assets, facilities and markets outside, as well as inside, the United States (Kilman 1998, B10). It has been predicted that further acquisitions will take place as other U.S. companies and European multinationals respond to this wave of consolidation. Between them, Monsanto and DuPont control 80 percent of the U.S. corn-seed market (Barboza 1999, S3, 1). DuPont has entered into a joint biotech venture with America's biggest seed company, Pioneer Hi-Bred International.

The Rural Advancement Fund International (RAFI) has warned that consolidation has "effectively marginalized the role of public sector plant breeding and research." The consolidation in the industry is motivated by a desire to extend control over patents, which RAFI regards as particularly pernicious for public research and farmers' rights: "The rapid formation of a seed oligopoly would be sufficient cause for government concern. But oligopoly hand-in-hand with intellectual property monopoly is a matter of grave concern. Even as the ranks of the seed industry implode, exclusive monopoly over varieties and genetic traits is exploding ... In combination with the global trade clout of the WTO, the global seed industry is positioning itself to dictate the future of plant breeding. When governments review the WTO's TRIPS provisions with respect to plants, they will be determining a key element in the fate of world food security"(1998a, 5).

Monsanto's acquisition of the cotton seed company Delta and Pine Land was of particular concern to rural advocacy groups. In conjunction with the U.S. Department of Agriculture (USDA), Delta and Pine Land had patented a technique for genetically altering seeds so that the resulting plant was sterile – its seeds would not germinate. The technique was dubbed "terminator technology" by RAFI, since it would prevent farmers from saving their seed to replant in the next season, a practice that occurs in both the developed and developing world but that is particularly critical for poor farmers in developing countries who are unable to afford new purchases each year. They not only ensure their survival by saving seeds but aid in the protection of biodiversity by maintaining traditional seed lines and developing and exchanging varieties within their communities. RAFI rejected arguments by the USDA and the company that farmers would not choose a technology that damaged their own interests. Although RAFI assumed that farmers were rational, it believed that the context in which their rationality was being exercised was becoming more and more limited by increasing consolidation in the seed industry (RAFI 1998b), and it was concerned that the terminator technology would extend a biological monopoly analogous to legal monopolies secured through patents.

However, Monsanto's takeover of Delta and Pine Land, a coproducer of terminator technology, was eventually abandoned in the wake of two challenges. First, antitrust regulators in the U.S. Department of Justice began investigating the proposed takeover, which would have given Monsanto control of 70 percent of the U.S. cotton seed market. The investigation was later expanded to examine the way that Monsanto's licensing arrangements for patented cotton genes could influence the market (Kilman 1999b, B12). Then in 1999, Monsanto CEO Robert Shapiro invited the president of the Rockefeller Foundation, Professor Gordon Conway, to address Monsanto's board of directors, expecting that Conway would give an encouraging speech on the virtues of genetically modified crops. Instead, Conway criticized the company's strategies and warned that the terminator technology's potentially devastating effects on subsistence farmers would not be outweighed by its benefits. In an open letter to Conway, Monsanto pledged not to pursue the technology (Vidal 1999), and the Delta and Pine Land deal fell through early in the new millenium, with Monsanto paying an $81-million termination fee ("Monsanto Pays" 2000, C4).

The company was also facing criticism by a coalition of U.S. farm and environmental groups, led by Jeremy Rifkin, head of the Foundation of Economic Trends and long-time biotechnology opponent. Their suit alleged that the safety of genetically modified crops had not been adequately tested and that patents gave the company too much control over staple foods (Kilman 1999a, A3). The spread of sentiments that fuelled the class-action suit had caused Europe's largest bank, Deutsche

Bank, to advise institutional investors to sell their shares in companies developing genetically modified organisms (GMOs), such as Monsanto and Novartis. In Europe, the consumer boycott of biotech foods had led to a drop in the price of genetically modified grains; the premium price for traditional varieties, Deutsche Bank warned, could outweigh any economic benefit from growing GMOs.

In late 1999 Monsanto began negotiating its own acquisition by Pharmacia-Upjohn, a merger that would create the world's eleventh-largest drug company. The *Wall Street Journal* reported that this signalled the end of Monsanto's biotech strategy: "the concept of a unified "life sciences" company – using technology to improve both medicines and foods – has become an affliction itself for Monsanto. The crop biotechnology half of the program has grown so controversial that Monsanto has agreed to a deal that is likely not only to push biotech to the back burner but also to cost Monsanto its independence" (Kilman and Burton 1999, A1). According to RAFI, Monsanto's acquisition was an inevitable result of its agricultural biotech strategy: "Faced with a huge debt load following $8.5 billion in agricultural input acquisitions; endless lawsuits claiming damages from its genetically-modified seeds; and plummeting share value and seed sales as producers and consumers back away from GM products, Monsanto had no choice but to seek protection in a larger enterprise" (2000, 1). RAFI warned, however, that the multinationals' life sciences strategy had not been abandoned completely, merely deferred.

Pharmacia is the senior partner in the merger with Monsanto, with its president and CEO becoming the head of the new company. Monsanto's agriculture business, which retains the Monsanto name, is separate, and approximately 20 percent was to be sold off (Kilman and Burton 1999, A1).[7]

MONSANTO AND RBGH

Monsanto perceived the marketing of rbGH as crucial to its biotechnology strategy. Although the choice of rbGH was somewhat arbitrary, having chosen the hormone, the company persisted with its development in the face of opposition from farm, consumer, and environmental groups, because of its significance as the first major product of agricultural biotechnology. It spent more than U.S.$1 billion on the product between 1982 and 1993 (Feder 1993, D4).

The history of the bovine growth hormone actually dates back to 1937, when Soviet scientists discovered that cows produced more milk when injected with pituitary-gland extracts (Asimov and Krouse 1937). The Soviets' results were confirmed by further studies conducted in England in the 1940s (Peel and Bauman 1987, 474).

In the 1960s, Monsanto attempted to synthesize a chemical version of the hormone, but this project was not successful and interest in bGH waned until the advent of recombinant technology. As Monsanto began its biotech trajectory in the late 1970s and early 1980s, rbGH was perceived as a landmark product, the approval of which would simultaneously revolutionize dairy production and advance the U.S. biotechnology industry. This perception was shared by the other three manufacturers of rbGH, Elanco, Upjohn, and American Cyanamid. In 1985, Monsanto predicted that the drug would be commercially available by 1988 and estimated the worldwide market for the product at $1 billion (Siler and Carey 1991, 74).

Industry spokespersons attributed significance to rbGH because of its status as a biotechnology product. In 1986 at the House of Representatives hearings into the impact of rbGH, the four manufacturers of the product spoke on each other's behalf about different aspects of its introduction. A spokesperson from Elanco, Edward Roberts, stressed the significance of rbGH for the biotechnology industry. He argued that the regulatory response to rbGH would send an important signal to the industry about the regulation of biotech products in general. The maintenance of the U.S. lead in biotech was perceived as critical to U.S. success in the marketplace, and as the first product of agricultural biotechnology, the approval of rbGH was seen as crucial to investor confidence in the industry (Roberts 1986, 82–84). The *Wall Street Journal* reported that the manufacturers saw themselves "[O]n the cusp of a genetic revolution. They see $500 million in annual sales if the Food and Drug Administration approves the hormone by the year's end, as expected. They also see it leading the way to a genetic make-over of the food system that will yield everything from leaner pigs to fatter walleyed pike" (Richards 1989, B1).

It was rbGH that enabled Monsanto to launch its biotechnology program. Since the company did not yet have an in-house pharmaceutical capability, rbGH was the first molecule available that it could scale-up to industrial production. After the denouement of a project to synthesize the growth hormone chemically in the 1960s, Monsanto had directed its development efforts toward plant genetics. When the start-up Genentech was formed in the 1970s, however, its first applications of recombinant technology were directed toward the development of human pharmaceuticals, rather than plant varieties. Genentech adapted the technology used to produce human growth hormones to animal hormones, and it became possible to create the molecule that Monsanto had earlier attempted to synthesize. Therefore, although Monsanto had an historic interest in the development of growth hormones, Monsanto spokespeople tended to emphasize the role of chance in the development of rbGH. A Monsanto official has noted that the choice of rbGH was somewhat arbitrary: "The science had not been developed to do this in plants, so there wasn't a plant program available. It was a matter of what was available, what wasn't, what had value and what didn't. Monsanto ended up with bST. In retrospect, we may have waited and chosen another molecule ... but hindsight's always 20/20."

Monsanto reached an agreement with Genentech to license the patent on rbGH and the process to ferment the recombinant organism. (To manufacture rbGH, the gene that produces the hormone in the cow is spliced into a bacterium, which is then able to produce the hormone. The genetically engineered bacterium is then grown under classical fermentation conditions, resulting in the production of more of the hormone as the bacteria reproduce. The rbGH is then separated from the bacterium and purified.) Monsanto's negotiations with Genentech leading up to the agreement illustrate the conflict between start-ups and multinationals over technology development. The value of a start-up company depends upon the strength of its technical know-how, and it is unwilling to give up the rights to its skills and techniques; the multinational, on the other hand, does not want to fund the development of a product that may compete with, or reduce demand for, its own product (Kenney 1986). According to a former Monsanto scientist, "Monsanto needed to be able to bring the technology in-house to acquire the necessary skills to scale it up to manufacture on a long-term basis. Although we had opportunities to allow Genentech to do part of that, it was our decision that from a long-term competitive standpoint, we needed to have those skills resident in Monsanto."

After the rights to the fermentation techniques had been licensed, Monsanto scientists began producing the material to see if it was effective. They then developed a prolonged-release delivery system for the hormone. Patenting was critical for Monsanto's interest in the technology. A former Monsanto scientist noted that "If you don't have the active patent, or some significant competitive barriers, it [the technology] rapidly becomes a commodity and most pharmaceutical companies have proprietary formulations; if they don't, they operate under an entirely different model, they can't really develop new technologies. The cost of development is too high, it doesn't allow you to develop something you don't have some protection over."

Investment decisions were made on the basis of the company's expectations about market needs and regulatory requirements. The two did not always coincide, however. A Monsanto spokesperson noted that the marketing department was focused on getting the product to market as quickly as possible in order to acquire the largest portion of market share; the scientists were more focused on obtaining regulatory approval. Monsanto's initial product was determined by its marketing strategy. The marketing department wanted a user-friendly product, that is, one that could be injected as easily and as infrequently as possible. Marketing personnel advised that farmers wanted a formulation that could be injected intramuscularly (I-M), like antibiotics. The FDA rejected the I-M route of administration, however, because of scarring of the muscle tissue, which may have affected meat quality if undetected during inspections (FDA interview). Subsequently, the company began trials with a product injected under the skin, or subcutaneously (SC). This lengthened the regulatory process. It was regarded by a Monsanto scientist as "One of those technical errors – the marketing people said farmers don't want to take time to do subcutaneous injections, they want to give it like antibiotics. It turned out this was a misperception ... that drove the whole thing."

The company had researched what formulation of the product would be acceptable to farmers but not whether the product itself was likely to be accepted. The literature demonstrated that milk yield could be increased significantly by injections of pituitary growth hormone, and it was therefore perceived that there would be a large market for the product. Monsanto scientists believed that the introduction of rbGH would have a major impact on the dairy industry and on agriculture generally. At the congressional hearings in 1986, the manager of the Agricultural Sciences Division, Lee Miller, discussed the huge increase in the productivity of American dairy farms and how rbGH could contribute to continuing productivity gains (Miller 1987, 107). The company did not foresee the reaction against the product by small farmers, consumers, and environmental groups in either the U.S. or in Europe. In fact, the rbGH manufacturing plant was built in Austria because it was expected that Monsanto would obtain approval in Europe before the product was approved in the United States, although, as it turned out, authorization in Europe has still not been granted.[8] Company managers and scientists explained that the European reaction was related to their concern about agricultural biotechnology in general and to a desire to maintain trade barriers against hormone-treated beef from the United States. Once the European Union had succeeded in banning hormone-treated beef, a Monsanto official stated, it was easier to reject other American biotechnology products.

At this time, the United States was experiencing unprecedented dairy overproduction, and farm income was falling. In 1988 a record 147 billion pounds of milk was produced in the country, with a surplus of 8 billion pounds. Since 1979, the government had spent almost $16 billion to buy and store surplus dairy produce (Schneider 1988, 45).[9] The surplus had been developing since 1965, as a result of a relatively constant 1.5 to 2.0 percent annual increase in milk yield per cow, without a corresponding increase in consumption of milk and milk products (OTA 1991b, 17). In 1981, the formula for the milk price support level changed. Since 1949, it had been based on a price that would give farmers the same purchasing power they had in a base period. But after 1981 the price mechanism was linked instead to the level of government purchases of milk and milk products. When purchases exceeded a certain threshold, prices fell; when purchases did not reach the threshold, prices rose (25). In 1985, Congress passed the Food Security Act, a bill that included provisions to reduce the U.S. milk surplus and its consequent drain on government revenues by instituting a whole-herd buyout program, in which thousands of animals were culled, and by reducing the price support for milk. In 1990 the government's price support level was reduced to $10.10 per hundredweight, where it was frozen until 1995 (25). These policies affected the traditional dairying regions differently from the way they affected the West and the South. The Pacific coast and Florida have large, industrialized production systems, with average herd sizes ranging from 500 to 1,500 animals and with the lowest production costs per unit in the country. Farms in the North and Northeast have herds of 50 to 150 animals and higher production costs; their share of dairy production fell by at least 2 percent during the 1980s. In 1988 cash income in the Upper Midwest fell below costs (20). Farmers in this region were finding it increasingly difficult to survive; the introduction of another production-enhancing technology would, some organizations believed, increase this pressure further and force small farmers out of business.

At this stage, Professor Dale Bauman, the principal investigator for trials at Cornell, was reporting increases in milk yield up to 40 percent from cows in experimental herds injected with rbGH (Bauman et al. 1985). An agricultural economist at Cornell, Professor Robert J. Kalter, used Bauman's data on production increases to make predictions about the effect of adopting rbGH technology across the industry. In his study, Kalter stated that productivity increases could reach as high as 25 percent in "well managed herds" and predicted that productivity increases of this magnitude would lead to rapid adoption of the technology. The study concluded that as production increased, milk prices would fall; therefore, "The number of dairymen and the size of the dairy herd will, by necessity, decline as the market seeks a new equilibrium" (Kalter et al. 1985, 117). At congressional hearings on the rbGH issue in 1986, Kalter observed that adoption of rbGH technology could cause a reduction in farm numbers and that without price supports there could be a loss of "15,000 farmers in New York State" (1986, 151). According to Kalter, the technology may favour larger operators, not because of the nature of the technology itself but because of the cost of additional technologies – such as computer-monitored feeding – that would ensure the success of the product (152).

Monsanto did not endorse this position. The manager of the company's Animal Sciences Division, Dr Lee Miller, argued that the "efficient" dairy farmer would be the primary beneficiary of the technology and that the efficiency of a farm was not related to its size but to how well it was managed (1987, 108). Spokespeople were puzzled by the negative reaction from small farmers, especially in Canada and Europe, because they perceived the technology as "size neutral"; that is, it did not require a huge capital outlay, unlike earlier dairy innovations such as milking parlours. The company did not forsee that the dairy surpluses of the 1980s would influence public acceptance of the product, believing that any product that increased efficiency would be well received by dairy farmers. A Monsanto official noted that "We didn't fully understand the impact. In the late '80s, there was an excess of cheese and butter products, there was storage of huge amounts of cheese, and we're talking about how exciting this production drug is! The first congressional hearing was in June 1986. At the time of the hearing they had a buyout program; Congress spent a huge amount of money to buy 10 percent of the dairy herd and to keep incentives for farmers not to go back into dairy ... then they heard there's a new product to increase milk production by 40 percent, so that caused some concern."

Monsanto even introduced a thirty-day billing method so that farmers would have higher milk receipts from increased production before they had to pay for their rbGH shipment. The company also offered farmers a discount on the product based on the percentage of the herd being treated, rather than on the number of animals (interview with Monsanto official). A Monsanto representative distinguished the "farming family" from the farm itself: "One of our advisors said, "I'm not interested in saving the family farm, I'm interested in saving the farming family. In other words, the family farm may not be the only unit the farming family wants to work with. They may want to farm but they don't want to farm their own farm, they want to work on someone else's farm. Or they may want a farm unit that's much bigger than what they had, that's 150 acres instead of 100. The farming family is what you want to help. If somebody wants to farm, we should try to help them do that."

However, what is suggested by this quote – that the concentration of farm ownership will increase – was a trend that concerned farm groups. Farmers recognized that the application of the technology might – at least initially – provide benefits to individual operators but argued that the resulting production increases would be damaging in the long run. A representative from the U.S. National Milk Producers' Federation stated that "More dairy farmers will be forced to adopt the new technology to remain competitive, only to see potential overproduction force down farm prices ... Gradually, the less efficient will not be able to survive and adjustments will be inevitable. That will hasten the day when technology will force those who remain to either get larger, become more efficient, or both. As you get larger you inevitably must consider new forms of management control and structure such as partnerships and incorporation. Gradually, the smaller, or less efficient dairy farm is forced out. The question we need to ask is, is this progress for America's agricultural system?" (Stemler 1987, 185–6).

Some farmers expressed their support for this conception of agriculture at the congressional hearings; they agreed that in the context of decreasing government support, the technology could help them to remain competitive (see VMAC 1993). However, others organized to counter this model of competitiveness. Protest against rbGH was expressed most vociferously in two dairy states, Wisconsin and Vermont. The situation for many farmers in these states was desperate, because milk prices had not kept pace with rising production costs, farmers were being driven out of business. In New York State, a farm family estimated that their return for seventeen-hour days was thirty-five cents an hour (Halpern 1988, 34). Farmers in Vermont and Wisconsin wondered whether their past adoption of technology had not, in fact, created the problem they were now confronting and began to question the viability of increasing productivity. In Vermont, farmer Stanley Christiansen, who in 1969 was known as "one of the most progressive and efficient farmers in Washington County," felt that "he spent a lifetime cutting his own throat" (Hiss 1994, 87). The director of the Dairy Forage Research Center in Madison, Wisconsin, contended that "We are in a period of relative luxury ... We can afford to take a hard look at our farming systems. We don't need to develop technologies that yield increases at any cost. We do need to introduce tools and policies that mean we'll be able to farm 1,000 or 2,000 years from now. What we're really talking about is a paradigm shift in our thinking about agriculture" (Schneider 1988, 47).

The concerns of farmers in Wisconsin and Vermont were confirmed with the release of a report from the Office of Technology Assessment entitled *Technology, Public Policy, and the Changing Structure of American Agriculture* (1986). The report advised that in order for American agriculture to remain internationally competitive, new technologies should be rapidly adopted; the OTA also acknowledged that adoption would reinforce the trend toward a dual-structure American agricultural system, in which farmers in the South and Southwest would be able to produce milk more cheaply and efficiently than those in the North and Northeast. According to the report, the different production levels across the country could be attributed to the quality of management and the philosophy and progressiveness of the farmer. Producers in the Northeast and Upper Midwest would need to change their farming practices in order to remain competitive with new production levels in the more "progressive" South and Southwest (7). The report stressed the need for careful management in order for rbGH to be effective: "Poor management results in a near zero response from bST supplement. Facets that contribute to the quality of management (and milk response to bST) include the herd health program, milking practices, nutrition program, and environmental conditions"(4). It also noted that "The ultimate gains to be captured depend not on the technology *per se,* but on the management skills of its adopters" (45).

The report identified bST as one of a series of biotechnologies that would revolutionize the farm industry and eventually render even bST itself obsolete. Included among these technologies were reproductive techniques that would enable the creation of transgenic cattle, new vaccines and diagnostic kits, and in food processing, techniques to improve the production of yoghurt and cheese (51).

Farmers' organizations such as Rural Vermont and the Wisconsin Family Farm Defense Fund opposed the drug by forming coalitions with other organizations, lobbying their state and federal congressional representatives, and advocating moratoria and labelling legislation at the state level. In 1986, the Wisconsin Family Farm Defense Fund joined with the Humane Society of the United States and Jeremy Rifkin's Foundation on Economic Trends to coordinate action opposing the drug. In 1987 the Foundation on Economic Trends petitioned the FDA to conduct studies of the safety and economic consequences of the drug, but the petition was rejected because the FDA does not have the mandate to consider the economic impacts of new technology and because safety data must be submitted by the sponsoring company.

MONSANTO AND THE SCIENTIFIC COMMUNITY

University Scientists and rbGH Research

Monsanto depended on scientists outside the company to conduct safety and efficacy trials, to publicize positive results from the trials, and to endorse the product. Most of the safety and efficacy trials were conducted at universities. As will be discussed in chapter 5, university and company scientists shared a similar perspective on animal health issues, which differed from the perspective of the regulators. This difference cannot be attributed to company pressure, however. As Kenney has noted, "Critics who are worried about outside influences on the research agenda focus on the individual and his [*sic*] project – not recognizing that the act of funding itself creates the agenda. One need not ask or force an investigator to do specific research – one need only fund the proper scientist to do what he wants to do (1986, 59)." In the rbGH case, scientists were funded to carry out particular kinds of research, but this agenda tended to fit with the scientists' own perceptions of the relationship between the hormone and animal health. In cases where scientists had a different perspective that they wished to invesitgate further, it has been reported that there was pressure not to speak out. This has tended to result in conflicts among and accusations against particular individuals. The problem goes beyond the rbGH case, however. It is now expected that university scientists will work in partnership with private-sector actors. Governments also rely on university scientists to provide advice on external panels. Although panelists are required to declare a conflict of interest, this expectation then narrows the field of available experts. As Mary Wiktorowicz observes, "[I]t is not uncommon for conflict of interest to arise, since there are often few experts in a specialized area" (2000, 13). Also, as discussed in Chapter 5, regulators may request the advice of university scientists on an informal basis. Again, Kenney notes that "The notion of a conflict of interest individualizes what is in actuality a systemic problem" (1986, 113).

In order to meet regulatory requirements for data collection, Monsanto conducted its safety and efficacy trials on rbGH at six universities in the United States: Cornell, Missouri, Arizona, Utah, Florida, and Vermont. In Canada, contract research was also conducted at the

MacDonald College campus of McGill University, which was also the site of some Elanco trials. The three other manufacturers also conducted their trials at various university campuses in Canada and the United States.[10] Other companies conducted trials at other Canadian campuses; American Cyanamid, for example, contracted its research to the University of Guelph.

In the United States, the colleges of agriculture at land-grant universities (LGUs) were established in every state and territory with the passage of the Morill Acts of 1862 and 1890.[11] Their original mission was to provide a practical education for citizens who could not afford higher education. This mission was further defined by subsequent acts, which prescribed a tripartite role in teaching, research, and extension.[12] In partnership with the states, the United States Department of Agriculture has funded technology transfer, agricultural research, and extension at the LGUs. These three functions are intended to be integrated so that technologies appropriate to public needs are developed and transferred to the local population and skills are taught to local producers and individuals. Cooperative extension is ideally a two-way process in which the needs of the local population are communicated to the agricultural colleges and its research and teaching programs are adjusted accordingly. The function of science, as originally envisioned, was to serve the public interest through the development and dissemination of new techniques and practices (National Research Council 1996, 87).

The colleges do not exist in isolation, however, but are connected to the agricultural system, as a recent National Research Council (1996) report has recognized. It noted that farming is now part of a global food and agricultural system and that the colleges must be responsive to this system: "An understanding of the complex needs and evolving characteristics of the food and agricultural system is a necessary condition for the continuing relevance of the land-grant [colleges]" (21). One characteristic of this system is that the private sector has a greater role in agricultural research and information and technology transfer. Under these circumstances, the role of public research and extension may seem to have been marginalized; however, the report also notes that "extension can help guarantee that information that influences public policy and private decisions regarding the food and agricultural system ... is widely accessible, accurate, and science-based" (88). This would suggest that, rather than transferring information and technology based on its own research, at least part of the LGUs' mandate is now to disseminate information based on private sector research and development. The promotion of rbGH would therefore fit within such a mandate.

The universities' role in the conduct of rbGH research highlighted the function of the academy in the contemporary agricultural system and led to conflict at the local level. At the 1986 congressional hearings, the Dean of the College of Agricultural Sciences at the University of Madison-Wisconsin expressed some ambivalence about the outcome of the introduction of the drug. In Wisconsin, academic scientists recognized that local farmers were confused and angered by the university's decision to proceed with rbGH research in spite of its apparent contradiction with the whole-herd buyout program and the financial pain experienced by farm communities. They decided, however, that although the changes induced by rbGH would disadvantage some farmers and processors, research was necessary in order to forestall further decline in the industry as foreign competitors proceeded with technological applications (Jorgensen 1987, 148–9).

The principal investigators at university sites claimed not to have a personal interest in the objectives of the trial; however, the trial did provide them with an opportunity to conduct their own research. At one Canadian university, provisions for additional funding for research of interest to the principal investigator were built into the contract; this amounted to between twenty and twenty-five thousand dollars per contract. Company funding also enabled Canadian university scientists to obtain matching funds from the provincial government for their own research, which was perceived as one way of coping with government cuts to research funding. Through this system, the researchers were able to pursue research that was not necessarily of interest to the company but that was made possible by funding initially supplied by it. The trials also enabled them to gain access to sufficient materials.

Universities also benefited financially from the company contracts. Costs related to the trial, such as labour costs, equipment costs, or veterinary bills were charged to the company, with an additional percentage of these costs charged as overhead.

The companies benefited in several ways from their arrangements with universities. The data collected from a university site was more likely to be accurate. Once a trial had been established at a particular campus, it was easier to conduct additional trials there, or it was easier for another company to begin trials there: the principal investigator was familiar with the procedure, and technicians had been trained. In some instances, it was possible to conduct a number of different studies at the one site; at Cornell, for example, milk from the animal trials could be used for studies of nutritional composition and human safety. Canadian trials also provided companies with data that could potentially be used in their marketing strategy.

University scientists did not regard the publication conditions specified in the contract as restrictive. A Canadian scientist operated with an "unrestricted contract, which means it [didn't] restrict publication." But there was "a 60-to-90-day grace period for the company to review the article and come back with comments. Those comments [didn't] have to be incorporated." One contract (not necessarily with Monsanto) specified that the data could be published only as a co-authored paper with the principal investigators from the other trial sites; other contracts permitted the research group to publish its own data separately. An American scientist said he did not recall any language in the contract that delayed publication and felt that the companies encouraged publication:

> In real life, it usually works in reverse. The companies are so anxious to have you publish the results that the [peer-review] system is too slow. When you submit a paper for review in a scientific journal, an eight-month delay is a relatively short delay to get through the review process, and the average would be more like a year and a half and sometimes two and a half years. The abstract is a year to three years ahead of the peer-reviewed publication. I've never run into a situation where the company has said you can't present this at a scientific meeting .... The main thing in the contract is if we were going to present something the company needs to have it in advance, mostly what they want to know is if they're going to read something in the newspaper about it later.

This scientist felt obligated to inform the company of his public announcements. "Even if it's not spelt out in the contract, as the investigator I always send things ahead of time."

A Canadian scientist who had contracted with another firm also noted that he had published every analysis of the trial data he felt was worthy of publication and that the company had not opposed any publication, including later publications by a colleague that suggested that the product may negatively affect reproduction in later lactations. During the trial period, however, he did feel restricted from discussing the data with scientists at other universities: "We had discussions with other scientists about the bST response but not about the experiments. In fact, there was some reluctance to talk about this because there were four companies trying to get the products tested and approved, so there was a certain amount of competition, and [the companies] didn't want a whole lot of information being spread around about the protocols or the levels of compound being used." This can be contrasted with the degree of interaction with the company: "Part of the contract responsibilities was to supply the company with data on a periodic basis, and we were in touch with them routinely. The people responsible for monitoring the trials for the company also visited, so there was considerable discussion, and the results were made available to them before they were published."

These comments lend support to Kenney's distinction between the disclosure of results and the disclosure of research. Although the companies did not attempt to stop publication, the awareness of the proprietary nature of the research and competition between companies limited discussion of research *procedures* among university scientists. As discussed in chapter 5, the experimental procedures themselves – and the company's adherence to these procedures as described in the protocol – were of particular concern to regulators.

One of the most outspoken critics of the drug, Dr David Kronfeld of Virginia Polytechnic, has claimed that adverse drug impacts have not been reported promptly by university scientists. According to Kronfeld, in 1963 he was invited to give two seminars at Monsanto and had suggested that they attempt to synthesize growth hormone chemically. Almost twenty years later, when Monsanto bought the rights to the recombinant bacteria from Genentech, he and his colleagues at the University of Pennsylvania were delighted because they felt that they were well connected with Monsanto and would therefore have a good chance of acquiring the contract for undertaking the animal effectiveness and safety trials. The initial agreement was that Monsanto would contract with the University of Pennsylvania, and Cornell University. Kronfeld had "grave misgivings" about the contract that was signed with Monsanto, however; he wished to conduct further examinations of animal safety than were required under the Monsanto protocol.

As it happened Kronfeld then went overseas on leave. On his return, the Monsanto contract had been terminated, and two of his colleagues had been employed by the company. Kronfeld and another colleague then "went after Cyanamid" to try to obtain the contract to conduct that company's studies. But after four months, Kronfeld claims, the colleague told Kronfeld that Cyanamid regarded him as a risk because he was too interested in the side effects of the drug, and that they would agree to have the research conducted at the University of Pennsylvania only if Kronfeld were excluded from involvement with the trials. At this point, Kronfeld was moved into administration; he responded by moving to Virginia Polytechnic.

Kronfeld has claimed that the negative effects of rbGH have not been promptly and accurately reported by the university researchers conducting safety and efficacy trials: "In 1987, I wrote a paper examining the first nine long-term trials and pointed out that there had been adverse effects in seven of them but they weren't being reported openly and this paper put me on the outside of people who were getting money from the companies and put me on the outside of people in the companies." Kronfeld claims that in conversations with friends and former colleagues who are now with Monsanto, he has been told that because of competition between the companies, they did not release data during the approval process: "They felt that as long as in the long run they made all the data available ... that would make good business sense and is acceptable business ethics." In Kronfeld's view, the publication of data that did not mention adverse animal health impacts increased pressure on regulatory agencies to approve the drug.

The Public Relations War: the Role of University Scientists and Health Professionals

Monsanto also relied on academic scientists and scientists from professional medical associations to endorse the product, particularly after the FDA began enforcing federal regulations banning the promotion of the product before approval. Scientists were not pressured by Monsanto; some came forward voluntarily, because they were concerned that their work was being misused by critics and because they feared for the future of the biotech revolution if this product – which had a wide margin of safety – was not approved. Supporting the product, therefore, was done in the long-term interest of biotechnology research and development. Nevertheless, the scientists' actions furthered Monsanto's interest and fulfilled its plans to communicate its message through third parties.

However, these scientists did not have all the data which had been provided to regulators. They were convinced by Monsanto's interpretation of the data, and their involvement allowed the company to get its message across and to defuse controversy.

Monsanto in fact perceived the controversy over rbGH as a consequence not of the unpopularity of the product but of the inadequacy of its public relations effort. It regarded the negative publicity as a problem of education; the opposition would die down once consumers were properly informed about the product. This required an internal, as well as an external, information campaign to inform Monsanto employees about why the company was proceeding with the product. According to one Monsanto representative, "we had just as much internal debate as external." The story Monsanto's scientists tell about the company's investment in rbGH is a story about the journey from an "age of innocence" in which negative response to the technology was completely unexpected, to a "new paradigm" in which the company was well aware of the need to change its public relations approach, obtain the support of respected scientific bodies, and enlist third parties to transmit information to consumers.

Authoritative science was regarded as necessary but not sufficient for the success of the product. "When you've got the science behind you, that you've got a safe and effective product, there's just nothing going on," said an official. However, the science itself was not enough; it also had to be communicated to the public. The company recognized that in order to convince farmers to use the product, it had to convince consumers to drink milk from their cows. However, U.S. federal regulations forbade the company from communicating its views on product safety until after it had been approved. The company therefore relied on third parties, such as university scientists and professional medical associations, to get its point of view across. These parties were often motivated by a desire to counter the misinformation disseminated by critics, rather than by a wish to support Monsanto. However, they nevertheless were persuaded that the product was beneficial, and their endorsement of its safety included an endorsement of the product itself. Once individuals within an association were convinced, efforts to persuade the whole organization followed.

In its public relations battle, Monsanto garnered the support of various professional groups, including the American Medical Association (AMA), the American Dietetic Association (ADA), and the American Pediatric Association (APA). The associations were particularly useful because they enabled the message to be spread to the constituency that Monsanto was most anxious about: the concerned consumer. In the words of one Monsanto representative, "We realized there's no way we could go to every consumer in the country; most of them don't care anyway. We need to get all the scientific communities out there and people who would potentially be called upon to know about it." The company recognized that favourable assessments by committees such as the FDA's Veterinary Medical Advisory Committee or the international body Codex were not useful unless consumers were aware of their content. "The average person doesn't know about the expert committee of Codex." Later comments by Monsanto representatives highlighted the usefulness of this resource; one said that "when the approval came ... consumers did call. They went to these professionals, and it ended there."

Not only did the involvement of the professional associations enable Monsanto's message to reach a greater number of people, it helped to defuse controversy and deflect attention away from the company. The companies created a list of medical professionals and academics who could be called upon to speak in place of Monsanto representatives. As a result of this strategy, "A couple of things happened; first of all, the issue usually died away, and there was no controversy, because having an activist and an academic person in an exchange doesn't elicit that much news. Whereas as soon as we step into the place, or another company does, it makes big news. That's part of the lesson we learned on how you can really force communication."

The support of third parties became even more critical after the FDA had directed Monsanto to stop promotional campaigns, including those by the Animal Health Institute (AHI), the trade association that had coordinated the manufacturers' public relations effort. In 1991 the Center for Veterinary Medicine (CVM) wrote to Monsanto warning the company against improper promotion of the drug in the preapproval period ("Monsanto Assailed" 1994, 8).[13] This was a turning point for the industry; Monsanto recognized that information would need to be distributed indirectly. Professional bodies were willing to participate in the effort to assuage public fears about rbGH. An official noted that "Even without us prompting, people were so outraged that they said if you can't go out there and talk, I'm going to. So you had people ... who said this isn't right, these people [opposition groups] are polluting science and scaremongering and I want to speak out. Well, a university professor standing up and talking about something has a lot more credibility than Monsanto or any company coming out and saying 'believe us.' So it actually worked in a much more beneficial way by having these third party people standing up and talking about it."

In March 1991 the AMA's Council of Scientific Affairs published a report on agricultural biotechnology that had been adopted at the association's general meeting in 1990. The aim of the report was to "Promote the education of the medical community, and dispel public misconceptions." Since physicians were the scientific resource most readily available to the public, the council argued, they should be informed about the safety of the product in order to be able to alleviate consumers' concerns (AMA Council on Scientific Affairs 1991, 1429).

A spokesperson for the American Medical Association was motivated to speak because he feared that criticism from animal rights groups and groups opposed to biotechnology were a threat to science. He believed that biotechnology held enormous potential for medicine: "We are right on the wave of the greatest scientific revolution that has ever occurred in the history of mankind" (Image Base 1993) and he was dismayed at the prospect that this revolution might be defeated by groups opposed to it: "Our feeling was that this particular biotech product [BST] was going to be the most well-researched product there is, and if you couldn't win the public relations battle to accept milk from BST-treated cows you didn't have a ghost of a chance to get the public to accept other things in the scientific pipeline. It was a test case, and we felt it was a must-win case because the company had already invested a lot of money in R&D by the time it became a public issue."

A member of the Canadian Dietetics Association was persuaded by the argument that the drug would encourage farmers to improve their herd management, possibly even reducing the level of herd disease. It was also believed that individuals should increase their level of milk consumption; if this advice was followed, the additional production resulting from rbGH would not be a problem.

Scientists were also persuaded by their trust in the judgment of company scientists and managers and in the judgment of university scientists conducting research on the product. A representative from the AMA made this comment about the ethical and scientific judgment of Monsanto's CEO, Richard Mahoney: "[Monsanto's] CEO is a very bright intellectual with very high ethical standards. He didn't like to lose. He wasn't going to lose for any reason except the science was no good. We had quite a bit of interaction with him." He believed that Mahoney's fortitude led to the eventual marketing of BST: "If you didn't have a CEO like the CEO of Monsanto at the time, and a company with enough resources, they would have given up on the product long before it came to be marketed." Representatives from the American Dietetic Association (ADA) and its Canadian counterpart also trusted the judgment of their colleagues who were engaged in biotechnology research. A representative from the ADA was persuaded by her colleagues' position that biotechnology was not qualitatively different from traditional plant or animal breeding techniques but merely accelerated the pace of genetic change. A Canadian nutritionist had confidence in the abilities of her colleagues who were researching rbGH at McGill, and her statements were informed by conversations with them, as well as by her review of the literature at that time. The AMA said that "We reviewed every scrap of paper that had been published in the scientific literature. We also had ongoing conversations with Monsanto and followed the kinds of experiments that were done there; we also had conversations with the FDA. We also had public and private debates with [opponents of the drug]."

Although the support of the associations was advantageous to Monsanto, association representatives appeared to deny that contact with the company provided the impetus for their actions: "Of course Monsanto was very pleased that we were doing this, and they were pleased with the outcome of our evaluations, but we never had a formal relationship with Monsanto. We didn't do the evaluation for them, nor did we get paid for it."

Internal opposition to bST within the AMA was overcome by communication. "If we did have a group that was uncertain or wavering, or moderately hostile to the use of bST, we'd go see them and describe everything from what the substance is to what it isn't and what tests have been done, and as I recall without exception these audiences came behind us and the position we'd taken." A representative said that the association agreed to endorse the FDA's approval of bST because "Their conclusion was consistent with our conclusions. Second, because we knew it would be controversial and we wanted to weigh in on the controversy. Third, we wanted to advise the public of what our scientific process was. And I'm sure there was probably a call from Monsanto saying are you people going to say anything. Since the FDA's decision agreed with our report, it was quite easy to put out a statement of support."

In addition to distributing information to professional medical associations, Monsanto distributed material to academics, dairy organizations, dairy processors, and the university extension services. In their communications with industry, academia, and the medical profession, company managers stressed that rbGH was a protein hormone and that it was digested like any other protein, that bovine growth hormone was already present in milk and levels did not increase after treatment with the recombinant product, and that it was not biologically active in humans.

As mentioned, the extension services at land-grant universities in the United States have traditionally been responsible for transferring knowledge from the academy to the state's farmers. While local farmers have therefore been traditionally served by animal scientists, consumers and food processors have been served by food scientists. At Cornell University a network of people produced information on rbGH for distribution to consumers, food processors, and health professionals. The food industry was another group targeted by extension services. Cheese factory managers were given information on the safety of rbGH in case consumers called their factory with questions. One scientist understood his role as enabling consumers to make an informed choice: "Our job is not to promote or to detract but to allow people to choose. We do a lot of that in extension – how to choose." The level of publicity generated by the extension services was proportional to public controversy about the product. "The number of questions you get fuels the activity level. If there's a newspaper article about a product ... there are a lot of questions to be answered."

Questions were generally answered in the form of fact sheets. Scientists at Cornell cooperated with others from the University of Illinois, Michigan State, and the University of California to produce a news release for the Council for Agricultural Science and Technology (CAST).[14] Articles and fact sheets were geared toward their target audience: it was assumed that consumers, for example, were not aware of the difference between steroid and protein hormones, so this difference was emphasized in consumer communications. Doctors, however, were assumed to understand the difference between the two and would automatically be aware that protein hormones are digested. The news release responded to each of the critics' claims, not only on human and animal health issues but on economic and environmental concerns. A university scientist, Dr Barbano, also co-authored an article for the Journal of the American Medical Association (JAMA). The article concluded that the FDA had answered all questions about the safety of milk from treated cows and added that "comments from health professionals can play an important role in reassuring the public about safety of milk and refuting misstatements or misconceptions about bST" (Daughaday and Barbano 1990, 1005).

Powell and Leiss (1996) have argued that risk decisions must be communicated to the public in order to avoid controversy and to dispel misinformation. By entering the fray, however, scientists were subjected to pressure from two directions. In the case of rbGH, this pressure was intensified because the controversy began before the evaluation of the product was complete. Although the academic scientists could speak about the results of their own trials, the "safety" of the product could be adequately judged only by an assessment of multiple trials: data that they could not have access to until all the trials had been completed and analyzed. Second, the academic scientists were expected not only to express their opinion to the public and the rest of the scientific community, but were also expected to maintain the confidentiality of the data. Since the company viewed the university scientist as playing an important role in the debate, the scientists' public education function cannot, in this instance, be clearly distinguished from their function in directing controversy away from the company. The academics' function as risk communicators was inevitably complicated by their constituents' knowledge of their relationship to the company. Although the scientists who worked on the product supported its efficacy and safety, scientists who did have doubts about the adequacy of the testing were pressured not to express these doubts publicly. A Canadian scientist felt

> Caught between a rock and a hard place. You have on one side the public, including the farmers, who don't trust you because you're under the auspices of the company. The company paid for this research. They also for some reason are attacking us for things that are illogical; how can I say that they're not going to get an increase in mastitis? I can't. I don't know if your cows are going to respond. Then on the other side you have the companies, who say don't raise any questions when you're giving a talk to somebody. Don't stand up there and say you should look a bit more at IGF-I, yes, IGF-I does increase a bit more in milk, don't say that. Well, why shouldn't I say that, that's what there is!

This scientist did, however, raise questions about IGF-I in an article. (Insulin-like growth factor I (IGF-I) is a hormone-like substance that is produced in response to growth hormone. The debate about IGF-I is explored in detail in chapter 5.)

Monsanto has used a similar strategy with its other biotechnology products. The company now distributes information to scientific bodies early in the development process, so that these organizations can defend the technologies. Although "revolutionary" language was important in inspiring the first wave of biotechnology investment and although recent reports seem to suggest it continues to be important, such rhetoric inhibited public acceptance; the company learned that the public no longer welcomes technological breakthroughs wholeheartedly. In response, a Monsanto representative has said that new technologies are now to be positioned as "evolutionary not revolutionary" and that the company has decided not to release information about products in the pipeline until they have been approved by regulatory agencies. After rbGH, the company intends to "let the regulatory process proceed before publicizing the benefits"; that is, that information about the benefits of a product will not be publicly released until it has been approved by regulatory agencies.

The distribution of information among professional associations made it difficult to find a specialist familiar with the relevant food safety issues who had not had some contact with Monsanto. This situation created not an individual, but a systemic problem. If, as I have argued, the conclusions drawn from the scientific evidence depended on the assumptions scientists brought to it, a fuller and more complex debate would have required a diversity of opinion within the scientific community – and an expression of that diversity.

SUMMARY AND CONCLUSION

In the 1970s the petroleum and petrochemical industries faced a crisis in profitability. These industries, which had supplied the agricultural chemicals that had contributed to increased agricultural productivity, began investing in biotechnology as a means of restoring profitability. They applied biotechnology with the intention not only of developing new products but of extending the marketability of earlier products such as herbicides. The development of products such as rbGH depended on the perception that biotechnology would form the basis of an alternative growth model and on state policies that protected intellectual property – both domestically and internationally – and enabled public institutions to claim proprietary rights. In the case of rbGH, development also depended on assumptions about farmers' needs and goals. Monsanto scientists and managers saw the impending technological revolution as a means to continue the agricultural trend of the previous thirty years in which technological applications had been used to increase productive efficiency. The company assumed that this goal was socially acceptable, although the farm crisis of the 1980s had undermined its legitimacy. When resistance to the technology grew, however, it persisted with the product because it perceived the importance of the biotechnology revolution itself to be important. As the resistance became increasingly focused on the health and safety aspects of the product, Monsanto looked to scientists outside the company to endorse the product and support its scientific legitimacy. The scientists involved often agreed to do this because of their belief in the product's importance for the biotechnology revolution. However, such a role added to pressure on those who had public and regulatory responsibilities and furthered the company's goals.

Since the approval of the drug in the United States, Monsanto has become the dominant player in agricultural biotechnology. It has restructured to focus on the application of biotechnology in agriculture, pharmaceuticals, and food, and it has divested itself of its chemical concerns (with the notable exception of agricultural chemicals). Its recent acquisition of seed companies has bolstered this strategy. However, the strategy itself has been called into question by a backlash in Europe that has undermined the value of shares in biotech companies. Monsanto's actions were motivated by a crisis in the prevailing industrial model, but opposition to the new model has hindered the company's plans to lead the biotechnology revolution and has led to its continued reliance on herbicides for revenue.

3 The Political Context in the United States

INTRODUCTION

In 1993 the Food and Drug Administration in the United States concluded that rbGH was safe for cows and for the humans who drink their milk; any indirect risk to human health from animal health problems could be managed within the existing milk monitoring system. The reasoning behind this decision is explored in chapter 5; the purpose of this chapter is to outline the context in which the FDA conducted its evaluation of rbGH.

During the period of the evaluation, the FDA was subject to two major pressures. The manufacturers of the product were eager to have it approved because, as I outlined in chapter 2, it was perceived as a harbinger for the industry and its regulation was accorded disproportionate significance. The sponsors of the drug urged Congress not to introduce legislation prohibiting or delaying it because of its importance to the biotechnology industry, which in turn was portrayed as crucial to America's economic future.

On the other hand, farm, consumer, and rural advocacy groups opposed the introduction of rbGH. The degree of controversy led the FDA to announce its human health decision before the approval of the product. Because political power is fragmented in the United States and because there are investigative bodies within both the legislative and executive branches of government, the critics of the drug were able to have the FDA's procedures and decisions reviewed by third parties. The political conflict did not, however, change the outcome of the decision.

The drug was approved in the United States because FDA scientists decided that the risks to animal health and any consequent risks to human health could be managed within the existing system. The efforts by advocacy groups to have this decision reexamined in public fora had no bearing on the decision itself.

Although there were several avenues in which the health and safety implications of the drug could be publicly re-examined, the fora for critiquing biotechnology itself had been limited by the decision not to create a separate body of law for the regulation of biotechnology products and processes.

THE U.S. REGULATORY SYSTEM

Analysts such as Sheila Jasanoff have argued that the nature and outcome of scientific debates are affected by the structure of the political system in which they take place: "[A] fundamental feature of political organization – the allocation of political authority among the three branches of government – heavily influences the form and intensity of scientific debates relating to risk" (1986, 5). Because political power is fragmented in the United States, the regulatory process is closely supervised by both Congress and the courts. Congressional committees may convene public hearings into regulatory matters, and congressional representatives may request that the investigative arm, the General Accounting Office (GAO), inquire into the conduct of the regulatory process. For example, "In 1978, the FDA took part in fifty-one hearings before twenty-four different Congressional committees and subcommittees, the GAO issued eight investigative reports on issues affected by the FDA, and the Agency responded to 4,463 written inquiries from Congress" (Brickmann, Jasanoff, and Ilgen 1985, 45). Congress also has sufficient resources to acquire external expertise on policy-relevant science.

Regulatory action also takes place under judicial oversight.[1] The Administrative Procedure Act of 1946 authorized the courts to overturn decisions not based on "substantial evidence" (Jasanoff 1995, 69). As a result of "the extraordinary judicialization of the American administrative process," Jasanoff states, "Agency rule-making has acquired many of the characteristics of a formal trial. As a result, individual citizens and citizen groups have unparalleled right to intervene in administrative proceedings, to question the expert judgements of government agencies, and ultimately to force changes in policy through litigation" (1986, 56).

In parliamentary systems, in contrast, the judiciary is more reluctant to invalidate regulatory actions. There are also fewer mechanisms for opening up the regulatory process to public participation. The Freedom of Information Act in Canada, for example, is more restrictive than its counterpart in the United States, and there is no Canadian or European equivalent of the GAO. Although commissions of inquiry may investigate government action, they are rarely established, and in parliamentary systems the legislature is much less likely to challenge regulatory action than in the United States. Jasanoff therefore characterizes the U.S. approach as "formal, open, adversarial, and confrontational," in contrast to the Canadian and European approach, which is "informal, confidential, consultative, and cooperative" (56).

The relative openness in the United States does not necessarily assist in the resolution of policy conflict, however, but rather prolongs the examination of evidence that cannot in itself solve a political dilemma. Habitual questioning by Congress also serves to undermine, rather than to bolster, public confidence in regulatory agencies (Brickmann, Jasanoff, and Ilgen 1985, 96). In the case of rbGH, the nature of the political system in the United States enabled critics to use the investigative branch of Congress, as well as the investigative offices within the executive branch, to review and critique the decision-making process. Advisory committee meetings were also open for public comment. However, it is important to distinguish mechanisms for review from the decision-making process itself. In the rbGH case, public hearings served to communicate and legitimize the FDA's findings to the public, but they did not influence the outcome. In order to understand the decision in this case, one needs to examine the assessment of the evidence before it was publicly discussed. Scientists needed to come to some conclusions about the product before submitting those conclusions for review.

BIOTECHNOLOGY: THE REGULATORY ENVIRONMENT IN THE 1980S

The industrial development of biotechnology was encouraged by policies that treated biotechnology food and drugs no differently from their traditional counterparts. The Reagan administration decided that, rather than creating a separate body of law to address the development of recombinant technology, biotechnology products should be regulated under existing statutes. In 1985 the Office of Science and Technology Policy announced the establishment of the Biotechnology Science Coordinating Council (BSCC), a group comprised of senior administrators from the FDA, EPA, and NSF that was intended to coordinate biotechnology policy across the various government departments (Kingsbury 1986, 50).[2] Sheila Jasanoff has argued that the BSCC was created by the White House in order "to seize control of biotechnology policy" and "to serve as a possible counterweight to possibly recalcitrant regulatory agencies" (1995b, 325).

In June 1986 the Office of Science and Technology Policy published its "Coordinated Framework for the Regulation of Biotechnology" in the *Federal Register*. The working group that developed the framework had concluded that "Existing laws as currently implemented would address regulatory needs adequately" (OSTP 1986, 23302). Therefore, "Existing statutes provide a basic network of agency jurisidiction over both research and products" (23303). Neither biotechnology processes nor the resulting products presented any novel risks; therefore they did not require any specific legislation (Jasanoff 1995b, 157). This meant that foods and drugs produced using recombinant technology would still be regulated by the Food and Drug Administration (FDA) under the Federal Food, Drugs, and Cosmetics Act (FFDCA).

On 29 May 1992 the FDA reaffirmed the intentions of the Coordinated Framework of 1986 and clarified its interpretation of how the FFDCA should be applied to foods produced using new technologies (primarily biotechnology). The FDA considered "existing statutory authority ... fully adequate to ensure safety regardless of process" (FDA 1992, 22989). According to the FDA's interpretation, "the regulatory status of a food, irrespective of the methods by which it is developed, is dependent upon objective characteristics of the food and the intended use of the food" (22984). A food should be regulated not according to the means or methods by which it is produced but according to the characteristics of the final product.[3]

In August 1990, President Bush approved the "Principles for the Regulatory Review of Biotechnology." According to these principles, regulations must: focus on the product's characteristics rather than its method of production; they must accommodate rapid advances in biotechnology; they must specify performance-based, rather than design-based, standards (that is, manufacturers should be able to meet the specified regulatory goal by using a number of different product designs rather than designing the product according to particular standards); and they must "minimize the regulatory burden while assuring the protection of public health and safety" (OSTP 1992, 6760).

The President's Council on Competitiveness reiterated the principle that the regulation of biotechnology should be decreased. In its report on National Biotechnology Policy, released in February 1991, the council stated that "The Administration has sought to eliminate unneeded regulatory burden for all phases of the development of new biotechnology products" (6761). Two months later, the council issued the *Fact Sheet on Critical Technologies*, which argued that regulation should be issued only when the benefit gained from the regulation exceeded the cost of imposing it. Voluntary private standards and disclosure should be relied on. Where necessary, regulations should also be based on "scientific risk assessment," and licensing should be carried out swiftly, based on criteria clearly defined in advance (6760).

The distinction between biotechnology products and processes was reiterated by FDA scientists, who had been dragged into the rbGH debate. During the period in which the drug was under review, Congress debated the ethical implications of animal patenting, and the biotechnology issue was once again on the public agenda. In attempting to distinguish the rbGH debate from the concurrent biotechnology controversy, FDA reviewers reproduced the distinction at the heart of the administration's biotechnology policy: rbGH was not an example of biotechnology. Only the process, not the product, was the result of recombinant techniques. In their view, a biotechnology product was defined as "transgenic"; that is, it was defined as an animal or plant that had foreign genetic material incorporated into its DNA. According to this definition, milk or meat from animals treated with rbGH was not a product of biotechnology; "milk and milk coming from treated animals is not really biotech food or transgenic food – the biotechnology only impacts how the drug is made, it's just a manufacturing process." Since discussions about biotechnology were proceeding during the FDA evaluations, an FDA official noted that "there was a tendency for people to consider bST in the same light. So the agency and the firm had to do some work to dissuade people from the perception that somehow bST resulted in transgenic food being consumed."

THE FDA, ANIMAL DRUG LAWS, AND THE REVIEW OF RBGH

The U.S. Food and Drug Administration "regulates 25 cents of every dollar spent by the American consumer, or about $1 trillion worth of goods and services annually ... [it] employs over 9,000 people and has a budget of $1 billion" ("In Defence of the FDA" 1995, 981). The FDA's actions, therefore, affect not only Americans' health and safety but a significant portion of their economic activity. Throughout its history, the FDA has sought to protect consumers without unduly damaging the interests of the drug, food, and cosmetics manufacturers whose products it regulates. In the 1960s, the thalidomide tragedy in Europe and Canada prompted Congress to institute legislative reform that expanded the powers of the agency in order "to strengthen the laws designed to keep unfit drugs off the market in the first instance and speed their removal should they reach the market" (United States Senate 1962, 2884). In the 1990s, Republican pressure in Congress has attempted to reverse this direction and speed drugs *to* the market rather than *from* it (see Goldberg 1996).

The FDA is responsible for regulating human and animal drugs as well as food products and cosmetics under the Federal Food, Drug, and Cosmetic Act (FFDCA). Amendments to the act in 1968 created a separate body of law for the regulation of animal drugs and led to the establishment of a centre for their evaluation, now known as the Center for Veterinary Medicine. Full authority to determine the human health implications of animal drugs was not granted to the center until 1983 (Lambert 1997, 277n).

The Animal Drug Review Process

In order to have an animal drug approved for commercial release in the United States, the manufacturer of the drug must lodge an application with the Center for Veterinary Medicine (CVM).[4] The primary reviewer coordinates the review of the drug and is the main contact with the sponsor of the drug. He or she is usually also responsible for part of the review. In the case of Monsanto's rbGH product, Posilac, the primary reviewer was part of a team of reviewers responsible for evaluating its efficacy and animal safety, along with other CVM animal scientists, veterinarians, and statisticians (personal communication with FDA official).

The sponsor of the drug must demonstrate that the product is effective – that is, that the product does what the sponsor claims – and that it is safe. In the case of food-producing animals, the product must be safe for the target animal and for humans who consume milk, meat, or other products from it, and the manufacture of the product must not be damaging to the environment (OMB 1994, 9).

To conduct safety and efficacy studies, the manufacturers must establish an Investigational New Animal Drug (INAD) application with the Center for Veterinary Medicine. Human food safety data must be submitted to the center, which then establishes whether drug residues present any risk to human health. The manufacturer must also provide a method for measuring the presence of the drug residues in food products from the animal. During the investigations, milk or meat from test animals cannot be released into the food supply until after a withdrawal period – that is, a period in which the food products are withheld from sale after treatment with the drug – so that no residues are present when the product is consumed. If, however, the food will not contain any drug residues, or if the residues are considered harmless, the scientists may allow the food to be sold without a withdrawal period, in which case the manufacturer does not have to provide a method of measurement. Initially, the FDA had set a withdrawal period of five days for milk and fifteen days for meat from cows treated with rbGH in 1985; after it concluded that rbGH posed no health risk to consumers, it ruled that no withdrawal period was necessary and that milk and meat from cows treated with rbGH in investigational trials could be sold to the public (OMB 1994, A1).

Before beginning a trial, the manufacturers must submit drug shipment notices indicating the location of the trial, the expected timeframe, the number and type of animals to be treated, the maximum dose, the duration of treatment, and whether the trial is a "pivotal" study. A pivotal study is one used by the agency to decide the safety and effectiveness of the product. The sponsoring company must identify whether a study is pivotal or nonpivotal when it submits a notice of drug shipment for each study to the CVM. The center is usually involved in designing pivotal studies. All data from a pivotal study are to be submitted to the center as part of a new animal drug application (NADA). A nonpivotal study, on the other hand, is run by the company to look at certain aspects of the drug that may not *directly* concern the CVM. The company must submit the results of nonpivotal studies, but only in the form of a study report; it does not have to submit all the data. According to FDA officials, if the results from nonpivotal studies are inconsistent with those from pivotal studies or suggest adverse effects not demonstrated in the pivotal studies, the FDA may request more data (OMB 1994, 16), and/or the study may be elevated to "pivotal" status. This occurred with a nonpivotal study of the Posilac formulation of rbST on injection-site reactions at the University of Vermont, in which a problem appeared that was not anticipated by the FDA, and, as a result, the study data were reclassified as pivotal. Only the summarized results from pivotal studies (including the Vermont study demonstrating severe injection-site reactions) were released in the Freedom of Information (FOI) summary; results from other, controversial studies at Vermont were not released because they were not pivotal for Monsanto's Posilac formulation of rbGH. The FDA attempts to inspect the site of every pivotal study; in the rbGH case they "came close" to inspecting every site, according to one representative.

When the drug manufacturer considers that it has all the data necessary for final approval, it submits a New Animal Drug Application (NADA) that includes the results of the investigational trials, a proposed label, and an environmental assessment of the manufacture and use of the product. The CVM reviews the data, asks the sponsor to correct any deficiencies, and approves the drug if the results meet its criteria

(Young 1987, 64). Under the Federal Food, Drug and Cosmetics Act, the FDA does not have a mandate to review the social and economic impacts of the introduction of a drug (OMB 1994, 10). Reviewers must respond to the sponsor no later than 180 days after submission, whether to request further data or to reject or approve the product.

The FDA and rbGH

By the late 1980s the FDA was subject to increasing pressure due to concern from the public and the dairy industry about the safety of the drug. Before 1990 human health concerns were nonspecific. Although some reports had mentioned animal health problems (see Schneider 1988), individuals and protest groups raising concerns about the drug did not have access to specific information from the human health trials on which to base their disquiet. Much of the public was not aware that milk or meat from animals used in investigational trials could be authorized by the FDA as safe for consumption before the drug had been approved. When safety concerns about rbGH were raised, therefore, there was a great deal of alarm about the safety of milk from test herds. By 1989 some reviewers of the Posilac submission were receiving two or three phone calls a day about rbGH. The FDA was also receiving numerous letters and congressional requests for information. Time that would normally be spent reviewing the product was taken up with fielding questions. At this stage, the agency decided to become more "proactive," to discuss the issues publicly rather than responding to individual inquiries, "so that we could get back to work."

It is important to recall the statutory requirement to respond to the sponsor no later than 180 days after submission; presumably the reviewers would have had even more difficulty meeting this requirement if they had chosen to respond to individual inquiries. For the first time, FDA officials began discussing not only the rbGH evaluation process, but the standard animal drug evaluation process, including food safety authorizations. This was an unusual position for FDA evaluators to be in. An official observed that "The Agency is not in the business of prospectively defending particular products or classes of products. Under normal circumstances, we can't even acknowledge that such products are under development because that's proprietary information. We would only get into that business if someone brought that to our attention and said we know you are reviewing this product X and we think it's unsafe. Then we're forced, in some cases, to defend ourselves by setting the record straight with the facts of the issue to overcome misunderstandings."

The FDA was under increasing pressure to release the corporations' human safety data, but because they were proprietary information, the raw data could not be released to the public without the companies' permission. Initially, the agency considered releasing a white paper, an agency report that summarized results of studies evaluating the human safety of rbST and that described the basis for the FDA's decision on human safety. Instead, the FDA decided that publishing in a peer-reviewed journal would give the report more credibility (interview with FDA officials). The corporations authorized publication of an article by two FDA scientists, Judith Juskevich, from the Division of Toxicology, and Greg Guyer, from the Division of Chemistry. Their summary of the FDA's review of the human health data was published in the journal *Science* in August 1990. As will be discussed in chapter 5, Juskevich and Guyer concluded that the use of rbGH in dairy cattle presented no increased health risk to consumers.

Dr Samuel Epstein, from the School of Public Health at the University of Illinois at Chicago, criticized the FDA's claims in an article published in a non-peer-reviewed publication, *International Journal of Health Services* (Gibbons 1990, 852). Epstein's major concerns were the potential effects of IGF-I, and antibiotic resistance and residues. These concerns were shared by the Consumers' Union, which released its analysis in December, two days before the Technology Assessment Conference of the National Institutes of Health (NIH) was convened to address the issue.

The NIH was asked to examine this issue in response to requests for a third-party review from groups associated with the dairy industry and, according to an FDA official, in response to public concern. The Technology Assessment Conference on Bovine Somatotropin ran from 5–7 December 1990. The NIH committee concluded that, based on the data it had been presented with, the use of rbGH did not present a public health risk. The committee did suggest, however, that further research should be conducted on the action of insulin-like growth factors and noted that it did not have sufficient evidence to draw a conclusion about the effect of the drug on the incidence of mastitis, a bacterial infection commonly treated with antibiotics (NIH 1991a).

Although the NIH supported the FDA's conclusions with regard to human health, both the FDA and university scientists' conclusions were called into question by adverse animal health reports that were leaked to the public. These reports undermined public confidence, because university and company scientists had previously emphasized the safety of the drug. In the fall of 1990 a researcher with the University of Vermont (UVM), Dr Marla Lyng, provided the farm advocacy group

Rural Vermont with photographs of deformed calves born to cows from the investigational trials. Dr Lyng stated that between August 1989 and August 1990, five severely deformed calves had been born to cows treated with the drug. She also provided copies of the herd computer health records and a list of cows treated with rbGH that had been analyzed by a consultant-veterinarian for the Vermont Senate and House committees (Christiansen 1995, 8).

The following year Rural Vermont released a report based on these findings at a joint press conference with the legislature (11). Representatives Ted Weiss and Bernard Sanders asked the FDA commissioner to compare the Rural Vermont data with that provided by Monsanto. According to Vermont state representative Andrew Christiansen, FDA commissioner Kessler's reply showed that only one of the UVM trials, the Jersey study, had been reviewed by the FDA; in that study nine out of twenty treated cows had developed mastitis, compared with two out of twenty controls; the calving rate was 100 percent in control animals and 85 percent in treated cows. One of the treated cows aborted (12). Commissioner Kessler noted, however, that the FDA did not accept Monsanto's analysis of this data and that the company would be resubmitting data from at least eleven field trials to conform to FDA protocols. This revelation diminished Rural Vermont's confidence in the earlier public pronouncements made by UVM scientists that rbGH was not causing adverse effects on animal health. According to Christiansen, "This contradicted years of testimony and public statements by the University of Vermont and Monsanto that there had been no adverse health effects at UVM. In 1989 UVM and Monsanto worked as a team to promote rbGH to farmers. Monsanto hired a UVM extension expert ... to run several meetings for farmers ... He showed a slide show that was produced by Monsanto. The message was that rbGH would increase feed efficiency and milk yield. It would not change the milk. It would not hurt the cow or affect the calf. It would help the family farm" (12).

In a press conference in March 1992, UVM scientists acknowledged that the incidence of mastitis had increased, and that two cows had given birth to severely deformed calves. Data from the UVM trial indicating mastitis incidence and injection-site reactions in the treatment group were published in the *Journal of Dairy Science* in December 1992. The authors also noted, however, that the cows in the treatment group had more infection than those in the control group before the experiment began and cautioned against drawing conclusions based on the small number of animals in the trial (Pell et al. 1992).

The report on birth defects by Rural Vermont did cause the FDA reviewers to reexamine the Vermont data. According to an FDA scientist, the FDA had received all the Monsanto data that had been submitted in support of the animal drug application for Posilac and had not encountered other reports of significantly increased birth defects "So that drew a question in our minds ... but we had to delve into it a little more and that required looking into our records for the other bST products [tested in Vermont by Monsanto] and the inspection reports." This reexamination revealed that some cows that had calves with birth defects were in fact control animals, and this revelation eliminated an apparent effect due to rbST treatment. An FDA representative could not explain the incidence of birth abnormalities reported in the Rural Vermont data: "I know from growing up on a dairy farm that in a certain year you'd get something you'd never seen before, things pop up which are unusual. Whether there was a specific reason for these things to come up at UVM, we don't know. It wasn't a huge number of animals, but those things are low frequency, so anytime you have them, it's weird – but you do get weird years."

Representative Bernard Sanders requested that the GAO begin an inquiry into the UVM trials, but according to Andrew Christiansen the GAO terminated its investigation after eighteen months because it could not obtain data from the UVM or Monsanto (1995, 16). Vermont legislators senator Howrigan and representative Starr wrote to the FDA expressing a desire to work with them on the rbGH data. Christiansen says that in his reply Dr Guest included the cow identification numbers from the UVM study and alleged that Rural Vermont's analysis was affected by identification errors. Senator Howrigan did not receive this letter; Monsanto representatives obtained a copy of it, however, and distributed it to national news media (13–14). Rural Vermont believed that the confidentiality of its data had been violated by the release and expressed its disquiet in a series of letters exchanged with the FDA (16).

What was disturbing to the members of the Vermont State Legislature and Rural Vermont was not merely the release of the data but the apparent contradiction between the data and statements about the safety of the drug. The individuals concerned attempted to reach a resolution of the issue with the FDA, and, likewise, the FDA reexamined the issue and attempted to clarify it. The agency could not publicly release information about the trials before approval, however. The issue was complicated because not all the Vermont trials tested Posilac, the formulation that was eventually marketed by Monsanto; nor were the Posilac trials pivotal, so the results were not released in the FOI summary, except for the results from one trial that had been elevated to pivotal status. Pivotal studies had been conducted at Vermont with other Monsanto rbGH formulations.

On 3 November 1989 veterinarian Dr Richard J. Burroughs, who had been involved with the animal health data review, was fired. He had worked for the agency since 1979 (Burroughs 1994, 6). The FDA alleged that Burroughs was incompetent; but Burroughs responded that he had been fired for criticizing the review process. In an interview with the *New York Times*, Burroughs said that he had been fired after a long dispute with his superior over how the corporations' data should be interpreted. He added that the director of the Center for Veterinary Medicine, Dr Guest, had been meeting too frequently with industry representatives and had been criticized the previous summer by his staff. In response to these claims, the chair of the House Agriculture Committee, Senator Patrick J. Leahy, announced that the General Accounting Office would investigate Dr Burroughs' allegations (Schneider 1990, A21).

The Critics and the Investigative Arm of Congress

The GAO investigated the review of rbGH and concluded that the FDA had addressed the critical review guidelines in its studies of the direct effects of the drug on human food safety, animal safety, and drug efficacy. What had not been addressed, however, were the *indirect* effects on human health from animal health problems, primarily mastitis, and the antibiotics used to treat mastitis that could then leave residues in the milk supply. The study recommended, therefore, that the FDA should not approve the commercial release of the hormone until its relationship to mastitis had been adequately assessed (GAO 1992c).

The GAO had not examined whether the level of antibiotic residues in milk and meat would increase as a result of the introduction of the drug; but it noted that there was already concern about the current level of antibiotic residue and the capacity of the milk monitoring system to manage it. The safety of American milk is monitored through a cooperative program established by the FDA, the states, and the industry. The FDA is responsible for supervising the states' monitoring activities, introducing new test methods, and recommending additional drugs to be monitored (GAO 1992b). Two years earlier, the GAO had questioned the FDA's effectiveness in monitoring the milk supply for antibiotic residues. It claimed that the survey methodologies used by the FDA were not adequate for determining the level of antibiotic residue, because the surveys were not statistically valid, nor did they test for "extra-label" drugs (drugs that had not been approved for the use to which they were being put but that were nonetheless believed to be commonly used by the dairy industry). In 1992, the GAO noted that milk monitoring had improved but was still not adequate to ensure public health. In April 1991 the FDA agreed to expand the number of drugs to be screened and to recommend new methods that the states and industry might use for screening them. The agency also began its own testing program to screen for twelve, rather than four, drugs; however, the GAO stated that there were eighty-two animal drugs that could leave residues in milk, sixty-four of which either are commonly used or may present a health risk when consumed (GAO 1992b, 3). The agency's own program to test for twelve drugs in milk was of limited value because the number of samples taken was small (GAO 1992b, 2–5).

Given the GAO's earlier conclusions about the existence of antibiotic residues and the inadequacy of the FDA's monitoring capacity, it is not surprising that the GAO argued that rbGH should not be approved until a consensus had been reached on the mastitis issue. In response to the GAO's concerns, the FDA held an open public hearing with its Veterinary Medicine Advisory Committee (VMAC). This hearing, however, was perceived by the reviewers as a means of legitimizing the agency's actions rather than allowing for public input into the decision-making process. An agency official said that "It was an agency decision to allow public comment as a way of assuring the public." At the meeting, the FDA also stressed improvements to the milk-monitoring system that had been introduced since 1991 and announced that further improvements would be made in the next month (Mitchell 1993). After hearing evidence from FDA scientists, company representatives, academics, and groups opposed to the drug – including Rural Vermont, the Humane Society of the United States, and the Center for Science in the Public Interest – the committee concluded that the risk to public health from increased antibiotic residues was manageable and that the product was therefore approvable (VMAC 1993, 5).

However, the critics of the drug felt that the manageability criteria were arbitrary and that their fears about the status of this category had not been adequately addressed. Anthony Pollina, an aide to Bernard Sanders and former director of Rural Vermont, requested a meeting with the FDA to have this conclusion explained. The agency brought approximately twenty people to the meeting, leaving him feeling overwhelmed: "So, how are you going to win that argument?" He also objected to the supposition that the risks would be managed by farmers and the monitoring system, although the ultimate responsibility was not specified: "They were saying that these are things we think we can take care of, whoever 'we' are." He was concerned that the guidelines for use would not be followed in practice. "Our question was, if the product is a risk when it's used wisely, what if it's used unwisely?"

Like the scientists and executives at Monsanto, rural advocacy groups saw rbGH as a precedent-setting case, with regulatory action sending a signal to the biotechnology industry. Just as this perception increased the corporation's desire to see the product approved, activists found it all the more imperative that a thorough review be conducted and, consequently, that human and animal health questions could be answered satisfactorily. Pollina said that "If in this case, the product represents a manageable risk, what does that say about other biotech products coming down the line?"

Activists were also concerned that the agency had not taken appropriate action in response to Monsanto's preapproval promotion of the drug – through the distribution of material, seminars at universities, and market research activities – which violated federal regulations. Federal law prohibits a drug manufacturer or others acting on the manufacturers' behalf from representing the drug as "safe and effective" until the regulatory review is completed and approval authorized (Office of the Inspector General 1994, 1). Violations of this regulation were problematic because "such actions could contribute to the public's misunderstanding about the product, provide the sponsor with an unfair competitive advantage, and unduly influence the FDA in its role in reviewing a new animal drug" (Office of the Inspector General 1991: 5). Congressional representative Sanders asked the Office of the Inspector General to investigate Monsanto's promotional activity between 1991 and 1994 and the agency's response. Between 1991 and 1993 Monsanto had organized several focus groups and paid farmers one hundred dollars to attend; sponsored the production of a video in cooperation with the American Medical Association, one copy of which was released prior to rbGH approval; and made a presentation at Louisiana State University (Office of the Inspector General 1994, 3).

The inspector general concluded that the Center for Veterinary Medicine had generally responded appropriately to Monsanto's activities. It had sent a warning letter to the company in January 1991 (in anticipation of the forthcoming report), and although it had not responded to Monsanto's actions since 1991, it did not believe that these actions represented violations of federal regulations. The Inspector General agreed with this interpretation in two of the activities but thought that the Louisiana seminar warranted attention. It also noted that existing regulations did not provide the centre with clear criteria for identifying promotional activities that required regulatory action and those that did not. Revisions of the regulations were recommended.

From Rural Vermont's perspective, Monsanto's promotional activities and the agency's failure to respond promptly and forcefully shed doubt on the autonomy of the FDA evaluation process. Anthony Pollina thought that the FDA should have taken much stronger action against the company. Pollina and Sanders discussed the issue with FDA representatives, but were unsatisfied with the outcome. Pollina said that

> You don't know what to do after a while because you feel like the deck is really stacked against you. If you're a member of Congress, you go to the FDA or the Inspector General who investigates the FDA, and if they say this is what's happened, this is the remedy, then where do you go? If the Federal government is so tied to the corporate agenda, where else do you turn? Congress wasn't going to pass a law mandating the labelling of bGH, that was out of the question, we knew ... that the industry undertook a strong lobbying effort to cut off any efforts for mandatory labelling. Monsanto started years ahead of time with their propaganda to convince people that the product was good, the product would work, and labelling would be unnecessary.

MILK LABELLING

When the FDA's review of the animal and human health data was complete, it considered the issue of whether milk from rbGH-treated cows should be labelled. Critics who had opposed the introduction of the drug advocated product labelling as a means to allow consumers to express their opposition – whatever its basis – by choosing not to purchase milk produced using the drug. But the FDA decided that it did not have a legal basis to require labelling of rbGH milk. It believed that there were no significant differences between milk from treated and untreated cows and that therefore the absence of a label for treated milk could not be regarded as "misleading." Under existing regulations, milk labelling could be considered false and misleading unless it was based on a material fact, such as a health or safety consideration or a change in the product's taste, texture, or other characteristics. The safety decision, therefore, also determined the labelling decision; there was little option for consumer choice once the decision that there was little significant difference between milk from treated and untreated cows.

The FDA's statement of 29 May 1992 reaffirming its commitment to the "Coordinated Framework for the Regulation of Biotechnology" also outlined its interpretation of the regulations regarding food labelling. Section 403(i) of 21 U.S.C. requires that the producer of a food describe it by its common or usual name. The producer must "reveal all facts that are material in light of representations made or suggested by labeling or with respect to consequences which may result from use" (343(a) 321(n); FDA 1992, 22991). The FDA concluded that consumers must be informed if a food produced by novel methods is so different from its traditional form that the usual or common name no longer applies to the food; likewise, consumers must be informed if the novel food presents safety or usage problems. The method of manufacture itself is not considered to be material information within the meaning of section 201(n).

Consumers, however, notified the FDA that they wished to be informed whether a plant or food had been developed using genetic engineering. The FDA responded to these concerns in another statement on 28 April 1993. It reiterated its position that historically it has limited its interpretation of "materiality" to the attributes of the food itself (FDA 1993c, 25838). Although several consumers had pointed out that the FDA had allowed process labelling for irradiated foods, the FDA contended that it had done so because of the characteristics of irradiated foods, not the irradiation process *per se*. Irradiation could cause changes in the organoleptic properties (taste, smell, texture, or colour) of finished foods and that these changes could be significant in light of consumers' perception of the foods as unprocessed (OMB 1994, 17).

On 6 and 7 May, the Food Advisory Committee and the Veterinary Medicine Advisory Committee held a joint meeting to discuss the issue, and it was decided that the consumers' interest in labelling was not sufficient to influence the FDA's ruling (OMB 1994, 19). Milk could be labelled voluntarily by producers not using rbGH on their herds. However, such a label would be permitted only if it were not misleading. Since natural bovine growth hormone is present in milk, the FDA reasoned that a "bST-free" label would not describe the product accurately. Such a label might also imply that the untreated product is safer. Therefore, claims that the milk is from untreated cows would be permitted only if the statement was put "in a proper context"; for example, the information that "no significant difference has been shown between milk derived from rbST-treated and non-rbST treated cows" would put the label in context and prevent it from misleading consumers.

Subsequently, a number of states issued guidelines regarding voluntary labelling. Centner and Lathrop comment that "The painstaking efforts taken by legislatures and regulatory officials in many states to regulate products derived from rbST-treated cows show an immense concern over the use of this new drug" (1997, 550). Although the FDA recommended guidelines for voluntary milk labelling, Vermont initially introduced mandatory labelling for milk products produced using bST. If the processor was unsure whether the drug had been used, he or she should err on the side of caution and label the product. Processors were also empowered to ask for an affidavit from farmers up to ninety days before they intended to begin using the drug. Anthony Pollina believed that it was important that those who used the drug should be responsible for notifying consumers of its use. Under voluntary labelling "The 92 percent of farmers who aren't using it have the burden of labelling instead of the 8 percent who are. Why should I, as a farmer, who's never used this product, now have to allow myself to be regulated and inspected and bear the expense of the label? There's all these things I have to do simply because I'm doing what I've always done." He noted that the state government and the Farm Bureau had argued that "if you labelled your milk bGH-free it implied your milk was better, therefore you could charge a premium for it." Pollina objected to this perspective: "Our response was you're saying milk without bGH is going to become like a specialty food, and yet it's the same as it was for hundreds of years ... so essentially bGH-free milk, which is natural milk, would become a niche product, which is exactly what's wrong with the way we relate to food in America – something that's real becomes special."

Vermont's mandatory labelling law was challenged by a coalition of large food producers *in International Dairy Foods v. Amestoy*.[5] The district court rejected the plaintiffs' case, but the Court of Appeals for the Second Circuit reversed the district court's judgment on the grounds that the First Amendment protects the right to silence as well as the right to speech and that the violation of this right was not justified by the state's intent to ensure consumers' right to know. In a dissenting opinion, Judge Leval argued that the majority had misinterpreted the state's intent, which was not to fulfill consumer curiosity but to address people's concerns about animal health, biotechnology, and the livelihood of small farmers. On 30 August the Vermont Attorney General agreed to stop enforcing the mandatory law and decided not to appeal, citing prohibitive legal costs. The mandatory law was replaced with legislation for voluntary labelling in April of 1997. Under these provisions, milk producers could declare by affidavit that they had not used rbGH, and the milk handler would in turn produce an affidavit for the commissioner of agriculture, food and markets, who was authorized to conduct random farm inspection to verify that rbGH was not being used (Centner and Lathrop 1997).

Although a number of states have permitted voluntary labeling, several others – Illinois, Nevada, and Texas – do not permit it. Conflicting labelling laws in different states have created discord. The Vermont ice-cream maker Ben & Jerry's, which owes much of its success to its "all natural" image, sued the state of Illinois and the city of Chicago after regulators threatened to remove ice cream with the rbGH-free label from the supermarket shelves. Ben & Jerry's label also included the FDA disclaimer, but any type of rbGH label was not permitted in this state. The company reached an out-of-court settlement with the state and the city which enabled its products to be sold freely in Illinois (Rural Vermont 1997b) and permitted producers in Illinois to voluntarily label their product. Because Chicago is a regional distribution centre, Ben & Jerry's needed to win the Illinois battle in order to ensure that their products would appear, labels intact, across the United States.

APPROVAL

On 5 November 1993 the FDA approved Monsanto's product Posilac (FDA 1993b, 59946). In the final ruling, it was also announced that approval would not have a significant impact on the human environment and that an environmental-impact statement was not required (59947).

Immediately after approval had been announced, Senator Russell D. Feingold of Wisconsin sponsored a moratorium that delayed the release of the product for ninety days (Schneider 1993, 1). During this period, the Office of Management and Budget produced its assessment of the FDA review and the impact of approval of the product. The OMB report reiterated earlier statements that the product posed no threats to human or animal health and that it would merely reinforce productivity changes already experienced by the dairy industry. It also added that although lower milk prices were expected to contribute to higher federal government dairy-price support costs, these costs would be offset by decreased costs for nutrition programs like food stamps and the Special Supplemental Food Program for Women, Infants and Children. Another factor considered significant by the OMB was the negative impact of a moratorium on the U.S. biotechnology industry; U.S. leadership in the industry and private investment in research and development would be enhanced by approval but hindered by postapproval regulation by the government (1994, iii–iv). The moratorium was lifted on 5 February 1994.

As a condition of the approval of the drug, the FDA had stipulated that Monsanto establish a post-approval monitoring program (PAMP). This program involved tracking the milk production and drug residues from treated herds in twenty-one dairy states for two years. After twelve months, the amount of milk discarded (because of drug residues) in the postapproval period would be compared with the preapproval period. Twenty-four commercial dairy herds would be monitored for mastitis, animal drug use, and resulting milk loss. Monsanto was also ordered to report all animal health complaints to the FDA every ninety days as part of a proactive system in which the company sought out reports of adverse experiences (Department of Health and Human Services 1993).

By October, the company estimated that 7 percent of dairy farms, or ten thousand farmers, had adopted the drug. Milk production was 3 percent higher than in September of the previous year, an increase that the Department of Agriculture attributed to introduction of the drug. Prices dropped by 10¢, to $12.70 per hundred pounds of milk. The *New York Times* reported that some dairy farmers were experiencing animal health problems after administering the drug on their farm: a dairy producer from New York had stopped using the drug and had sold thirty-four of his two hundred cows after they had developed mastitis; another farmer in New York reported similar problems (Schneider 1994a, 11). In late summer 1994 the Wisconsin Farmers' Union and the National Farmers' Union, based in Denver, Colorado, set up a toll-free hotline to record information from farmers who had experienced problems with rbGH. A number of farmers reported problems that had led them to cull cows treated with the drug. The Farmers' Union investigated whether Monsanto had reported similar problems to the FDA. It obtained access to Monsanto's report by filing a Freedom of Information Act (FOIA) request and found that it also included incidents of death, outbreaks of mastitis, spontaneous abortions, and other health problems. The Farmers' Union observed that sixty-eight of the ninety-six reports had been forwarded to the FDA on 1 September; however, the FDA had told the Farmers' Union that any serious problem or any adverse reaction not listed on the label should be reported to the agency no less than fifteen days after the manufacturer became aware of the problem. In October the agency reported that Monsanto was conveying adverse reaction reports immediately (Kastel 1995, 3–9).

Monsanto, meanwhile, had been proclaiming the success of its product and referred to its marketing program as "unparalleled in the agricultural industry." Federal Express delivered the product directly from Monsanto to the farmer's door within forty-eight hours of ordering. Along with the first order of the drug, farmers received a $150 voucher to pay for a veterinarian's assessment of their herd (Monsanto 1994a, 1). In a supplement to its quarterly report, Monsanto cited USDA estimates that the product would be adopted by farmers for use in 10 to 15 percent of the U.S. dairy herd within a year of commercial release. With this adoption rate, Monsanto expected its animal science division would break even in 1994 and become an income contributor in 1995 (Monsanto 1994a, 1). In its 1995 annual report, Monsanto announced that Posilac "has already become the world's best-selling veterinary product to dairy producers." But in spite of this "It isn't yet profitable because of unsatisfactory manufacturing costs worsened by currency translation (it's made in Austria). We expect Posilac to become profitable this year as sales growth and improved manufacturing bring unit costs down" (Monsanto 1995a,3). By 31 January 1995 Monsanto had sold 14.5 million doses of Posilac to thirteen thousand dairy farmers; 2.7 million cows had been injected with the drug.

Monsanto stopped releasing Posilac sales figures in February, although spokespeople for the company claimed that there was still "steady growth" in the level of Posilac usage. But a survey of farmers undertaken by Rockwood Research during the summer of 1995 found that although twenty percent of farmers had tried the product, 87 percent of the farmers who had not tried Posilac said they would not use it in the future. Then in October, Monsanto introduced a 10 percent discount plan for farmers who purchased a six-month supply of the drug (Stayer 1995, 1, 8).

At the end of November 1996 the Veterinary Medicine Advisory Committee heard the final report on the findings from the Post-Approval Monitoring Program (PAMP). FDA officials stated that there had been no increase in the amount of milk discarded due to antibiotic residues since the start of rbGH sales. The evaluation of twenty-eight commercial herds concluded that the experience with the drug reflected most of the predictions on the label: mastitis was increased, foot and leg injuries were higher, and the use of medications had increased. Although some of the reproductive problems in the pretreatment studies were not found in the commercial herds and the incidence of mastitis was lower than originally determined (VMAC 1996), the regime for assessing mastitis incidence was less rigorous than in the university studies. The FDA said that it had inspected farms in New York and Florida – presumably those that had been reported in the media as having significant animal health problems as a result of the drug – and had discovered that the problems "related more to farm management practices than to the use of Posilac" (VMAC 1996: 29). But the animal-health critic David Kronfeld argued that these kinds of judgments actually inhibited the reporting of adverse drug experiences: "This has worked a bit against reporting because when a farmer reports that he's got a mastitis problem, Monsanto approaches him. They have an animal scientist and veterinarian and tell him to look at the label, the FDA agrees with us, it's management which is responsible for the mastitis, it's not the drug itself. This must be very discouraging for the farmer, and, in fact, I've had a number of farmers call me and I've talked with other people who have said that farmers are being discouraged from reporting adverse effects by this vigorous, enthusiastic, proactive program which will try to pin the disease on the farmer's management rather than on the drug" (VMAC 1996, 112).

Although the drug had been approved, questions continued to be raised about the integrity of the process and the FDA's links to the drug industry. In April 1994 Representatives George E. Brown Jr, David R. Obey, and Bernard Sanders asked the GAO to investigate the role of three officials in the approval process. The representatives argued that these officials had ties with Monsanto that conflicted with their role as evaluators of Monsanto's application for drug approval. The officials were Michael R. Taylor, deputy commissioner for policy, who had joined the agency in 1991 after working for the law firm King and Spalding, which represents Monsanto; Dr Margaret A. Miller, deputy director of the agency's office of new animal drugs, who was a former Monsanto employee; and Dr Suzanne Sechen, the primary reviewer, who had worked as a graduate student for Professor Dale Bauman, who conducted the Cornell trials on rbGH for Monsanto (Schwartz 1994, A3).

The GAO concluded that none of the three officials had conflicting financial interests and that their role at the FDA did not transgress Office of Government Ethics standards regarding the appearance of loss of impartiality. However, the ethics standards that applied to these officials had changed in 1993. Before 1993, FDA employees who had worked for a company were required to refrain from regulatory work involving that company for one year after commencing employment with the agency and prohibited for life from working on any issue or product they had had direct involvement with at that company. In 1993 the lifetime prohibition was lifted, and employees were to refrain from regulatory actions involving their previous employer for one year.

The GAO found that Dr Miller had not influenced the outcome of the health and safety review. However, she had helped draft the FDA's answer to a petition from the Foundation on Economic Trends seeking to ban the sale of milk and meat from cows treated in investigational trials, and she had signed the Freedom of Information document on human health safety. The GAO did not regard either of these actions as problematic, because they did not bear on the health and safety review itself, although the office noted that Miller's signature constituted a "technical violation." In 1993 Dr Miller had been asked by senior FDA officials to brief the commissioner, Dr David Kessler, on issues relating to the drug. It was reported that when Dr Kessler heard of Dr Miller's prior affiliation with Monsanto, he asked that her conduct be investigated, but she had not violated the FDA's standards of conduct or the

Office of Government Ethics standards of conduct (GAO 1994: 15). The GAO did, however, identify several articles authored or co-authored by Dr Sechen and Dr Miller, some of which had been written with the FDA listed as their address, "whose publication may have been contrary to FDA's requirements for prior approval of outside activities" (GAO 1994, 1). Articles written by Dr Miller had been coauthored with Monsanto scientists; articles authored by Dr Sechen had been coauthored with, among others, Professor Dale Bauman at Cornell University, who was principal investigator of trials of Posilac conducted at Cornell.

Michael Taylor, like Dr Miller, had also been required to avoid Monsanto-related regulatory action for one year. He had also advised Dr Kessler that he would not be involved with policy decisions relating to the Posilac application. However, in 1993 Taylor had signed the guidance on the voluntary labeling of milk and milk products from treated cows (20). He had also taken part in discussions during the drafting process, but according to other FDA employees he had not sought to influence the content of the guidance. Since by the time these guidelines were drafted he had been with the agency for over a year, no regulations had been violated. With regard to Dr Sechen, the GAO concluded that she could not have a conflict of interest because she had never been a Monsanto employee.

The director of Rural Vermont was unhappy with the GAO's conclusions; he believed the officials who had been investigated "clearly" had a conflict of interest, either because of their employment history or because they had conducted research on Monsanto's product. He was particularly disturbed by the implication that "the FDA didn't know when their employees were doing official duties and when they were doing outside activities, which doesn't give you a lot of confidence." He regarded Dr. Miller's discussions with Commissioner Kessler as problematic because the issues on which her opinion was sought – such as the biological significance of the increases in mastitis and the likelihood of antibiotic contamination of milk – were central to the approval of the product.

In addition to requesting investigation of the FDA decision-making process, consumers also used the judicial system to challenge the FDA's approval of rbGH and its failure to impose a mandatory labelling requirement on milk and other dairy produce made from the milk of treated cows. In *Stauber v. Shalala,* plaintiffs argued that the FDA's approval was arbitrary and capricious because it had not fully considered the implications of either increased mastitis incidence or higher milk IGF-I concentrations because it had not requested an environmental impact statement, and because it had failed to require mandatory labelling of products derived from treated animals.[6] The court dismissed these arguments. With regard to the failure to request an environmental impact statement, the court noted that after it had reviewed an environmental assessment from Monsanto that concluded that Posilac would not have a significant impact on the environment within the meaning of the National Environmental Policy Act (NEPA), the FDA had decided that such a statement was not necessary. An environmental impact statement is not required if the environmental assessment determines that no impact will occur. A statement concerning the socioeconomic impact of the introduction of the drug can be required only if an environmental impact statement is also necessary, that is, only if there is evidence of potential environmental harm. Therefore, the court denied all of the plaintiffs' arguments on the labelling issue, responding that labelling can be required only if there is a material difference – that is, a difference in the taste, smell, or shelf-life of the product; consumer interest is relevant only if a material difference can be detected.

SUMMARY AND CONCLUSION

The Reagan administration fostered the growth of the biotechnology industry through its changes to patent and tax law, its refusal to regulate the products of biotechnology separately, and its creation of a biotechnology coordinating committee that could be used to rein in skeptical regulatory agencies. But because of – or perhaps in spite of – the administration's attempt to ease the regulatory burden on the industry, the first major product of agricultural biotechnology encountered opposition from the public and within congress that slowed the regulatory process.

In their eagerness to promote both the drug and the industry, the manufacturers publicized its benefits widely. When adverse reports came to light, however, critics, farmers, and rural advocacy groups were concerned not only about the data themselves but about the apparent contradiction between them and public statements made by the companies and university scientists. The symbolic importance of the product meant that advocacy groups, as well as corporations, regarded the signals sent out by the regulatory decision on rbGH as critical for the future development of the industry. The contradiction led to demands for investigations of the process and caused the FDA to defend its evaluation before it had been completed, a highly unusual situation for the agency.

Given the nature of the U.S. political system, the critics were able to have the FDA's actions investigated by the GAO and other parties.

Congressional representatives also pursued conflict of interest allegations against FDA employees with an association with Monsanto. The GAO exonerated the employees; however, critics were not reassured, since regulations imposing a lifetime ban on FDA staff working with products with which they had previously been associated had been altered. The regulatory changes reflected a systemic reality of greater interaction between corporations, universities, and regulatory bodies.

The critics raised important questions about the data and procedures for regulatory review. However, they were not able to affect the interpretation of the data that ultimately determined the decision or the interpretation of the labelling law which prevented mandatory action and imposed the burden of notification on those not administering the drug.

4 The Political Context in Canada

INTRODUCTION

In Canada, the Bureau of Veterinary Drugs within Health Canada rejected the drug rbGH on animal health grounds. Canadian regulators, like their U.S. counterparts, found that the drug did not represent a threat to human safety. However, scientists within the bureau who had not been involved in the human health review questioned this decision. As will be explored further in chapter 5, these scientists were concerned that there was insufficient evidence on which to base the human health decision and that some of the data contradicted the conclusions on which the decision had been based. The internal – and external – controversy about the product continued for nine years and was resolved only when the Canadian government asked two external panels appointed by the Royal College of Physicians and Surgeons to review the decision. The external panels supported the conclusion of the Bureau of Veterinary Drugs in the case of both human and animal health. However, the panel decisions have not ended problems within the bureau. The review process for animal drugs continues to be fraught with conflict, and scientists have continued to allege that they are under pressure to approve animal drugs that they believe may potentially have a negative impact on people consuming meat and milk.

As in the United States, a number of farm, food policy, and citizens' groups had requested investigations into the impact of the product and the review process. Although there were fewer avenues open to these groups in Canada than in the United States, the drug received an unusual degree of attention. In terms of the length of the review process and the scrutiny to which it was subjected, the review of rbGH was not typical of the Canadian system and was somewhat closer to the system in the United States.

The decision was made in the context of a Canadian dairy system governed by supply management and of a legal and financial regime that provides less support for the biotechnology industry than does the regime in the United States. Although the Canadian government has stated its support for biotechnology and although certain policies have been changed to fulfill this goal, the dramatic policy changes that fostered the biotechnology industry in the United States have not been copied as aggressively in Canada, potentially placing less pressure on scientists in the Canadian context. Scientists claimed, however, to have been directly pressured by the company. But whatever the substance of these allegations, it did not influence their assessment of the data which ultimately proved decisive.

BIOTECHNOLOGY IN CANADA

In 1998, the Canadian biotechnology industry generated almost $2 billion in revenue, including $750 million in exports (Industry Canada Bio-Industries Branch 1998, 2). The Canadian government has implemented several measures to encourage the development of this sector. It introduced a Canadian biotechnology strategy in 1983, invested in biotechnology companies through Technology Partnerships Canada, and supported Genomics research through the National Research Council and the Medical Research Council (Manley 1998). In order to comply with international agreements for the protection of intellectual property, Canada has increased protection for brand-name drugs at the expense of generics, thereby encouraging the expansion of the biopharmaceutical industry. However, Canada has not encouraged the growth of the biotechnology industry to the extent that the United States has done. During the period of the rbGH review, patent protection, funding for basic science, and incentives to commercialize university research were significantly lower than in the United States.

Canadian Patent Law

The Canadian intellectual property rights regime was challenged by the introduction of the North American Free Trade Agreement (NAFTA) which came into effect on 1 January 1994, and by the negotiations for the Uruguay Round of the General Agreement on Tariff and Trade (GATT). The Uruguay Round negotiations resulted in the creation of a World Trade Organization (WTO) and an Agreement on Trade-Related Aspects of Intellectual Property Rights (TRIPS) that came into force on 1 January 1995 (NAFTA Secretariat 1996, 10).

As a result of the TRIPS and NAFTA negotiations, Canada's intellectual property regime – which differentiated between drugs produced in Canada and drugs produced elsewhere – was altered to provide increased protection for brand-name pharmaceutical companies. In 1993, in order to meet Canada's obligations under the agreements that were then being negotiated, Bill C-91 was introduced. The new bill was intended to promote the development of the drug industry in Canada, while still ensuring that drugs were available at "nonexcessive" prices (Manley 1997). Bill C-91 eliminated compulsory licensing, under which generic versions of patented drugs could be produced in Canada seven years after the brand-name manufacturer's product had been approved and could be imported ten years after brand-name approval. Compulsory licenses were not granted for drugs that had been developed in Canada (Food and Drug Law Group, 1994, 325). The compulsory licensing provisions had helped contain drug prices and had fostered the development of the Canadian generic drug industry.

Article 1703 of NAFTA enforces the principle of nondiscrimination and national treatment in the protection and enforcement of intellectual property rights; the protection provided to domestic claimants must also be extended to other signatories (333). Compulsory licenses may be granted, but only under certain restrictions.[1] Under both the WTO agreement and NAFTA, the same level of intellectual property protection is to be provided across sectors; Canada's differential protection for the drug industry contravened this obligation.

In addition to eliminating compulsory licensing, Bill C-91 introduced regulations that linked the issuance of a notice of compliance (regulatory approval) with the patent status of the drug. For example, if a brand-name manufacturer contested the validity of a generic competitor's patent claim, Health Canada would not issue a notice of compliance to the generic manufacturer until the patent issues had been resolved, or a certain time period had elapsed.[2] The confidentiality of data submitted by drug manufacturers was also protected by Bill C-91. Paragraph 5 of article 1711 of NAFTA requires that data submitted to a regulatory agency to determine the safety and effectiveness of a product must not be disclosed by the agency, unless the disclosure is necessary to protect the public or the data is protected from unfair commercial use (Food and Drug Law Group 1994, 337).

In order to limit price increases due to extended patent protection for brand-name drug manufacturers, the legislation strengthened the powers of the Patented Medicine Prices Review Board (PMPRB) and allowed exceptions to patent infringement for regulatory approval and stockpiling (Manley 1997).

Since 1993, the Canadian biotechnology industry has expanded dramatically, a development that the industry attributes to Bill C-91; between 1993 and 1997 the number of Canadian biotechnology companies doubled, and by 1997 the industry employed three thousand more people than the generic companies (McKenna 1997, B4).[3]

However, elements of patent law have remained unchanged under TRIPS and NAFTA provisions. Although microorganisms must be patentable, signatories are not required to provide patent protection for higher life forms (animals and plants – although some form of intellectual property protection must be provided for plants). Certain provisions enable parties to the agreement to reject patent applications if a proprietary claim threatens human, animal or plant life, or the "*ordre public* or morality" (see NAFTA article 1702(9)).

Under the Canadian Patent Act, an invention must be "new" and "inventive" in order to be considered patentable (McMahon 1995, 24). Unicellular living organisms have been patentable since the 1982 Patent Appeal Board ruling in the Abitibi case (which predated TRIPS and NAFTA). The board found that since the organism had been created by an inventive step, was useful, and had not existed previously in nature, it constituted patentable subject matter (14). Higher life-forms, including plants and animals, were not patentable in Canada at the time of the rbGH review. An application to patent the "Harvard," or "onco," mouse was rejected by the commissioner of patents in August 1995.[4] The application had been accepted in the United States in 1988 and in Europe in 1992 (McMahon 1995, 19).

Unlike other major trading partners, Canada does not offer patent term restoration, in which some of the patent time lost in the clinical trial and regulatory review period is restored to the patent owner; nor does Canada have a procedure for opposing patents once they have been issued (NBAC 1998, 51).

Biotechnology Policy in Canada

The development of a Canadian biotechnology industry has been encouraged by a system of government grants that have "incubated biotechnology companies in their infancy" (Scoffield 1998, B6). In 1983 the government created a national biotechnology strategy that was designed, among other things, to attract investment to Canada and focus biotechnology R&D in areas of "strategic importance" (Envision Research 1997, 11). The national biotechnology strategy is administered by the National Biotechnology Advisory Council (NBAC) and an Interdepartmental Committee on Biotechnology (ICB). However, the OECD has noted that "neither NBAC nor the ICB had sufficient mandate nor resources to assume an effective 'proactive' role" and that funding levels were inadequate to build a coherent strategy, support research at companies and universities, and encourage co-ordination between the two" (Intellectual Property Policy Directorate 1998, 2).

Biotechnology had been one of the six sectors slated for regulatory improvement in the government's report *Building a More Innovative Economy*. The government had defined biotechnology as an "enabling technology" that could transform the basis on which industries can compete (Industry Canada 1994, 63). According to the report, because the biotechnology industry had specified regulatory uncertainty as the main obstacle to the development of the industry in Canada, action was required to ensure continued investment in this country (30). A number of regulations, including those pursuant to the Fertilizers Act, Seeds Act, and Food and Drugs Act, were revised by the end of 1996.

In the Budget of 1995, the government set out a strategy according to which future science and technology research "would be concentrated more strategically on activities that foster innovation, rapid commercialization, and value-added production" (Minister of Finance 1996, 74). To achieve this goal, it announced in 1996 that it was reallocating $270 million to technology-development programs over the next three years. One such program, Technology Partnerships Canada, funds private sector investment in new technologies. Areas funded by the program have included biotechnology and other "enabling" technologies such as advanced manufacturing and materials technologies, aerospace, and defense conversion. With existing resources from Industry Canada, the technology partnerships program had funding of $250 million by 1998–99. A technology network to promote technology diffusion was launched, and the current system of tax incentives for scientific research and experimental development was evaluated (75–6).

The Canadian government also supported the biotechnology industry through its funding of basic research by organizations such as the Medical Research Council (MRC).[5] According to the president of the MRC, two-thirds of Canadian biotechnology companies had been started after receiving seed money from the organization, including BioChem Pharma, the manufacturer of the world's best-selling AIDS drug. The money available through this source had diminished by the late 1990s, however: MRC funding dropped to 1992 levels in 1997–98 (Scoffield 1998, B1). The National Biotechnology Advisory Committee (NBAC 1998) noted that "Canada's once world-class biotechnology science base is eroding under cuts in public funding" (32). In the United States, the NIH budget was $10.7 billion in 1997, having risen 16.3 percent in the previous three years. In the same period, the MRC budget declined by 10.7 percent to $238 million, representing 2.2 percent of the NIH budget (35). Although venture capital funding has increased since 1994, investment has gone toward development rather than basic research (Scoffield 1998, B6).

The commercialization of university research is another area in which Canada differs from the United States. Legislative changes in the United States in the early 1980s simplified and consolidated the patent regime, granted universities patenting rights over their researchers' inventions, and obligated them to transfer the results of their research to the private sector. In Canada, however, universities have been free to develop their own policies with regard to the commercialization of research.[6]

Biotechnology Regulation in Canada

Biotechnology is defined in the Canadian Environmental Protection Act (CEPA) as "the application of science and engineering in the direct and indirect use of living organisms or parts or products of living organisms in their natural or modified forms" (CEPA s.2 (3)(1)).

Biotechnology in Canada is regulated under a framework very similar to that applied in the United States; that is, biotech products are not treated differently than other products, and until the creation of the Canadian Food Inspection Agency (CFIA), which took over the inspection functions of several government departments, existing departments had been charged with their oversight. However, rather than allowing for greater regulation of biotechnology, the CFIA has been criticized for its dual mandate and for its failure to enforce its policies. Environmental groups and the House of Commons Standing Committee on Environment and Sustainable Development have lobbied for the creation of a separate statute and regulatory body for the oversight of biotechnology products; however, their recommendations have not been implemented by the Canadian government.

In 1992 Cabinet agreed to adopt common principles for the regulation of biotechnology products, principles that had been formulated by nine federal departments, including Health Canada and Environment Canada, and led by the minister of agriculture and

AgriFood. The three main principles were: "the recognition of the primacy of health and safety standards; the use of existing legislation and existing institutions to administer them; pre-release assessment of the risks involved in releasing organisms into the environment" (Morrissey 1995, 80).

On 11 January 1993 the federal government announced that these principles would form a regulatory framework for biotechnology. As in the U.S. Coordinated Framework, biotechnology products would be regulated by government departments under existing legislation.

Under this proposal, the introduction of biotechnology products would be administered by different departments. Biological pesticides, for example, would be regulated under the Pest Control Products Act, transgenic seeds under the Seeds Act, and pharmaceutical drugs – including rbGH – under the Food and Drugs Act, which is the responsibility of Health Canada. Any products not covered by existing statutes would be regulated under part 2 of the CEPA, "Toxic Substances." In 1997, the inspection functions previously carried out by the Departments of Agriculture and Agri-Food, Fisheries and Oceans, Health Canada, and Industry Canada were taken over by the Canadian Food Inspection Agency (CFIA). The CFIA regulates genetically modified crops and assesses their environmental impact (Royal Society of Canada 2001, 35), but Health Canada is still responsible for human food safety.[7]

Environmental groups, particularly the Canadian Institute for Environmental Law and Policy (CIELAP), had argued for more stringent regulation of biotechnology through the creation of a separate statute and regulatory body. The House of Commons Standing Committee on Environment and Sustainable Development endorsed CIELAP's proposal recommending that the Canadian Environmental Protection Act be amended to include a new part to deal specifically with the products of biotechnology.[8] The new part would include minimum notification and assessment standards for all biotechnology releases into the environment; other statutes and regulations would prevail only if their standards were equivalent to or greater than those in CEPA (House of Commons Standing Committee on Environment and Sustainable Development 1995, 124).

The government did not accept the committee's recommendations and maintained CEPA as a safety net for biotechnology products that were not adequately covered by regulations administered in other departments. A spokeperson from CIELAP criticized the proposed changes to CEPA as weakening existing law relating to biotechnology (Matas 1995, A1, A2).

In 1997 Industry Canada was charged with the task of revitalizing the National Biotechnology Strategy. The creation of a publicly acceptable ethical framework to guide regulatory policy was regarded as important for the successful renewal of the strategy. With this end in mind, Industry Canada made two hundred thousand dollars available for research into the regulation of biotechnology in other jurisdictions, and Health Canada established a series of round table discussions across the country to obtain public input on regulatory policy. Although the Round Table discussions were purportedly intended to obtain public input on biotechnology policy, the subsequent reports noted that lack of advance notice meant that many interested parties were unable to attend (Government of Canada 1998b, 3).

In 2001, an expert panel appointed by the Royal Society criticized the Canadian regulatory regime governing biotechnology, stating that "It appears that no formal criteria or decision-making framework exists for food safety approvals of GM products by Health Canada" (37). The assessment of genetically modified crops had been guided by the concept of "substantial equivalence"; that is, that if the product was found to be essentially the same as its unmodified counterpart, it was regarded as safe. The panel argued that the application of this principle prevented a rigorous safety assessment and that substantial equivalence need to be proved rather than assumed.

DAIRY FARMING IN CANADA

Canada's dairy farmers have traditionally been protected from the impact of oversupply and international competition by a supply-management system that regulates the quantities of milk produced and levies high tariffs on dairy imports. In Canada there was a debate about the role of rbGH in a dairy system that would be increasingly subjected to competition not only from the world market but from U.S. and other suppliers who had already implemented rbGH. Company representatives emphasized the threat to an unprotected Canadian system if rbGH was not introduced; Canadian farmers, however, argued that the drug would only add to the pressures of international competition. They were particularly concerned about the impact on profitability of a drop in milk consumption by Canadian consumers, who had contacted farm organizations in record numbers in order to express their opposition to the drug.

Jurisdiction over Canadian agriculture is divided between the federal and provincial governments. The responsibility for administering dairy price supports was taken over by the Canadian Dairy Commission (CDC) in 1966 (Skogstad 1987, 47). In 1969 the Dairy Farmers of Canada (DFC) instituted a plan that placed quotas on industrial milk and cream (used for the production of foods such as cheese or yoghurt). Farmers were eligible for a federal government subsidy only on goods produced within a quota allocated by the provincial milk board. Since 1975 the CDC has administered the quota system in conjunction with the Canadian Milk Supply Management Committee (CMSMC), which represents each provincial marketing board (103–4).

Import restrictions were a crucial part of maintaining the system, and Canada had imposed quotas on the import of dairy goods. Under the GATT provisions of the Uruguay Round, quotas were no longer permitted as a form of import restriction and were replaced by tariffs of up to 351 percent. The U.S. government protested that tariffs on poultry and dairy goods conflicted with Canada's obligation under NAFTA to eliminate tariffs by 1 June 1998. However, a five-member trade panel ruled in Canada's favour on 15 July 1996 thus allowing Canada to maintain its high border tariffs indefinitely (Fagan 1996, A8).

Although protected from international competition, Canadian dairies have undergone significant structural adjustment due to technological change. The adoption of technology has resulted in a near-doubling of milk yield per cow and has led to fewer and larger dairy farms. In 1965 the average herd size was fourteen cows (Stonehouse 1987, 5); by 1996, it had risen to fifty-six (Canadian Dairy Network, personal correspondence). An analysis in 1990 suggested that this tendency would be reinforced with the adoption of rbGH: "The high degree of managerial skill required to profitably utilize BST, combined with early adoption of this technology by certain producers, would encourage high-cost producers to leave the dairy industry. Early adopters of BST, facing decreases in cow numbers ... could be expected to purchase more quota to ensure full utilization of fixed resources. This is likely to accelerate the ongoing rationalization process of fewer but larger dairy farms with higher yields per cow" (Stennes, Barichello, and Graham 1990, 78).

In recent years, the Canadian government has introduced changes to agricultural policy. In 1991 the Agricultural Stabilization Act was repealed and replaced with the Farm Income Protection Act (FIPA). In the 1992 budget, the government announced a 10 percent decrease in the dairy subsidy effective from 1993. A 15 percent decrease was announced in the 1995 budget (*Canada Gazette* 1995), followed by a further 15 percent decrease in 1996. In the 1996 budget, the government announced that from August 1997 the subsidy would be phased out entirely over five years (Department of Finance 1996, 41).

RBGH IN CANADA

The drug rbGH is regulated under the Food and Drugs Act, which is administered by Health Canada. Initially, two companies, Monsanto and the Provel division of Eli Lilly, applied for regulatory approval, or a notice of compliance (NOC), in Canada. Monsanto submitted an application for a product under the trade name Nutrilac in 1990. Provel asked for its submission to be placed on hold in May 1996, pending the outcome of Monsanto's submission. The evaluation of Monsanto's product was undertaken by the Bureau of Veterinary Drugs within Health Canada, which approved the sale of milk and meat from test trials in 1988 (Senate of Canada 1998).

As in the United States, the manufacturer of the drug must submit information that includes a description of the drug; its brand name; a list of ingredients; a description of the plant and equipment to be used in manufacturing; the method of manufacture; details of tests to determine its potency, purity, stability, and safety; reports on its safety under conditions of use; substantial evidence of its clinical effectiveness for the purpose indicated; names and qualifications of the investigators to whom the drug has been sold; and a draft of the label to be used (Health Canada 1995a, 3). The studies to determine safety and efficacy include toxicity studies, pharmacology and residues studies, and animal safety studies (Foster 1994, 47). If the bureau is not satisfied that the information provided is sufficient, it can request additional data from the manufacturer. Once the review has been completed and the drug has met the requirements under section C.08.002 of the Food and Drug Regulations, the regulations stipulate that the minister of health shall issue a notice of compliance (Health Canada 1995, 3).

On 7 March 1994 the Standing Committee on Agriculture and AgriFood commenced hearings on the impact of the introduction of the drug in Canada. The hearings provided a forum for the discussion of the impact on dairy farming in Canada and also for industry groups to defend the scientific legitimacy of their safety claims. A number of departments and organizations were represented at the hearings, including Agriculture and AgriFood, Health Canada, Monsanto and Eli Lilly, the Animal Health Institute, the Dairy Farmers of Canada, the National Dairy Council, the Consumers' Association of Canada, and the Canadian Consumers' Association. A representative from the U.S. Consumer Policy Institute, Dr Michael Hansen, and the veterinarian Dr Richard Burroughs, who claimed that he had been fired from the FDA for criticizing the review process, also spoke at the hearings. The committee also received briefs from over sixty organizations and individuals.

Spokespeople from the Dairy Farmers of Canada and the National Dairy Council expressed concern about consumer reaction to milk from rbGH cows. The president of the Dairy Farmers of Canada, Peter Oosterhoff, requested a 180-day moratorium if an NOC was issued, so that consumers could be educated about the safety of the drug and so that there would not be an adverse consumer reaction (Oosterhoff 1994, 22).The National Dairy Council president, Kempton Matte, requested a two-year delay in approval and/or use of the drug in Canada (Matte 1994, 4).

Members from the Canadian Animal Health Institute (CAHI) also spoke at the hearings. Based in Guelph, Ontario, the institute is registered as a trade organization and is funded by more than thirty chemical, pharmaceutical, and biotechnology companies, including Monsanto and Eli Lilly. It produces and distributes scientific literature on products manufactured by these companies. According to its executive director, Jean Szkotnicki, promoting rbGH had been the institute's biggest program in recent years (Saunders 1995b, A13).

At the hearings, scientists from CAHI and Monsanto emphasized the legitimacy of the scientific evidence that verified the safety of the drug. Szkotnicki asserted that the manufacturers of rbGH had released their world-wide body of human safety data for publication in an article in *Science*; its conclusions had been verified by a number of respected organizations that had found that the consumption of products containing rbGH was safe for humans. She also noted that a scientific panel of the Canadian Medical Association, the Canadian Pediatric Society, the Canadian Veterinary Medical Association (CVMA), and the University of Toronto Faculty of Medicine had found the drug to be safe, as had the FAO/WHO Joint Expert Committee on Food Additives (Szkotnicki 1994, 24). Statements by the (CVMA) later became controversial when the organization was asked by Health Canada to strike an expert panel to review the Bureau of Veterinary Drug's evaluation of the product.

After the hearings, the standing committee produced a report that was released in April. It argued that a number of outstanding issues needed to be investigated further. One was that the Canadian dairy supply management system was already under strain, since one of the pillars of the system, import quotas, had been removed with the implementation of the GATT. The introduction of rbST, the committee argued, would put additional stress on the system. It also pointed out that because the Canadian dairy system is different from that of the United States, American economic-impact studies could not accurately predict the effects in Canada, nor could studies that ignored changes affecting the supply management system. (A study by Deloitte and Touche for the Canadian Animal Health Institute was cited as an example of this.) The risk of anitbiotic residues in milk, which was of great concern in the United States, was not regarded as a problem in Canada, because of stringent antibiotic testing. However, animal health concerns, particularly mastitis, were still an issue. An adverse consumer reaction was regarded as critical to the Canadian industry. Again, the committee believed that tracking the U.S. consumer reaction would not necessarily prefigure the Canadian response. Nor had the impact on Canada's export of breeding animals, semen and embryos – valued at $100 million – been determined. Therefore, the committee recommended

1 that the government should legislate a one-year moratorium on the use of rbGH in Canada;
2 that this period should be used to review the impact of its introduction, including the issues described above;
3 that a government-industry task force should be struck to undertake this task;
4 that imported dairy products should be labelled to demonstrate their conformity with the Canadian moratorium;
5 that mechanisms should be implemented to ensure greater transparency in the regulatory system;
6 that Health Canada and Agriculture and Agri-Food Canada should establish consistent procedures for handling biotechnology products; and
7 that the government should make provisions for assessing the likely socioeconomic and environmental effects of these products.

The government responded to the recommendations in August. Although it claimed to endorse them, it actually claimed that most of them could not be fulfilled because of either domestic or international statutory obligations. Rather than a legislated moratorium, the government had obtained a voluntary delay from the manufacturers until 1 July 1995. It endorsed the committee's second and third recommendations for an impact study and established a task force to carry out the review. The government also said that it endorsed recommendation 4; however, in its interpretation such labelling was implied by existing requirements that imported dairy products must show their country of origin. Countries that had licensed rbST were listed in an appendix to the report. It also noted that all milk contains trace amounts of natural bST, which is indistinguishable from the synthetically produced version; that no country had required mandatory milk labelling; and that the FDA guidelines did not permit labels to state that milk is bST-free or that it is safer than milk from untreated cows.

Although the government endorsed the recommendation for greater transparency in the decision-making process, its response was simply to maintain the existing process. It said that it would continue to provide information while respecting the limitations on disclosure imposed by the confidentiality provisions of the Canadian Access to Information Act, article 1711(5) of NAFTA, and article 39(3) of the GATT. Recommendation 6 was also endorsed; however, this endorsement did not signify any change in the government's approach to biotechnology. It noted that regulations and guidelines for agricultural products were discussed at a multistakeholder forum held in November 1993. Development of the regulatory process was continuing; Agriculture and Agri-Food Canada and Health and Industry Canada would develop consistent procedures for handling biotechnology products. Recommendation 7 was partly endorsed. It was pointed out that risk-based environmental safety assessment was already an accepted component of the regulatory review process. Furthermore, "Assessment of the possible socio-economic effects of biotechnology products is not supported as a regulatory criterion because these factors could pre-empt decisions based on safety, and effectiveness. The standard procedure in Canada and other industrialized countries is to regulate products based on scientific principles. Products are assessed for safety and effectiveness. Once safety and effectiveness have been reviewed, it is the marketplace in Canada which then decides on the market acceptance of the product, based on benefits such as price and individual values and preferences"(Government of Canada 1994, 8).

A number of environmental, citizen, and food policy groups had organized against the introduction of rbGH and had critiqued the government's response to the product. The Toronto Food Policy Council (TFPC) was among these groups. It was established in December 1990 by Toronto City Council, following a recommendation of the *Healthy Toronto 2000* report. The TFPC operates as a subcommittee of the city's board of health, and its members include city councillors and volunteers from business, farm, and consumer, and labour groups, as well as from multicultural, anti-hunger advocacy, faith, and community development groups. In its examination of rbGH, the council concluded

> that the approval of rbGH would represent a very significant failure in the Canadian food and agriculture policy making system. This failure stems primarily from a drug review process that does not require consideration of issues such as long-term public health implications (in this case, consumer acceptance of dairy products), and the impacts on the structure of the dairy industry and dairy farmers. Nor does the review process begin with the most basic questions: what problem is rbGH designed to solve? Is there a problem with the quality of the Canadian milk supply? Do we have a milk shortage in this country? Is milk production inefficient? (Toronto Food Policy Council 1995, 11)

The Food Policy Council wrote a letter to the minister of Agriculture and Agri-Food, Ralph Goodale, criticising the government's response. It applauded the government's implementation of a voluntary moratorium and its support for a system of biotechnology review, but it did not believe that the government's response to recommendations 4, 5, and 7 constituted an endorsement of them or that the task force appointed to examine the issue could "do anything but support the government's desire to approve the product" (Toronto Food Policy Council 1994b).[9]

The Food Policy Council made a number of recommendations that were subsequently adopted by the Toronto board of health: that the board of health should inform the federal and provincial governments that it did not believe that rbGH should be approved, that in the event of licensing of rbGH, dairy suppliers should indicate whether rbGH-produced milk was being sold to the schools, and that parents be so advised by the school boards (City of Toronto Board of Health 1994a, 3). A letter informing them of the council's action was sent to the federal and provincial health and agriculture ministers and to the chairman of the Standing Committee on Agriculture.

Meanwhile, the government had put together a task force to examine the issues suggested by the committee. The task force included representatives from Eli Lilly and Monsanto, the Canadian government, and the dairy industry, along with a consumer representative.[10] Since the task force was a fact-finding body, it made no recommendations to the government. During its terms, the members of the task force traveled to New York state and interviewed a scientist and veterinarian who had conducted research for and worked with Monsanto.

An examination of the economic impact of rbST adoption was undertaken by the Policy Branch of Agriculture and Agri-Food Canada. It findings depended on the proportion of farmers adopting the drug, the cows treated, the length of treatment, the yield increase per cow, the impact on costs of production, and the consumer reaction. While noting that adoption of the drug need not necessarily increase over-quota production levels in Canada, since it could be used as a management tool to improve the efficiency of production at existing levels rather than increasing production, it also pointed out possible adverse effects if the public reduced its consumption of milk because of concerns about the drug. Furthermore, a 3-percent decline in fluid milk consumption could reduce farm profitability by approximately 2.4 percent (Agriculture and AgriFood Canada 1995, i–v).

Four scientists reviewed the impact of rbST adoption on genetic evaluation and improvement programs in Canada (Dekkers et al. 1995, i–ii). Genetic improvement programs rely on the integrity of evaluation programs that record an animal's milk production levels and her genealogy. These programs have been in place in Canada since 1881: 65 percent of Canadian farmers participate in genetic evaluation programs, and 75 percent make use of the records produced by them (Daniel 1996, 1). Dekkers et al. stated that the introduction of rbST would reduce the accuracy of genetic evaluations, because once injected with bST, the cow's production level may no longer be indicative of her genetic capacity. They predicted that this could reduce rates of genetic improvement by 3 to 7 percent, and recommended that the use of rbST in Canada be recorded in order to minimize the impact on genetic evaluation. They also stated, however, that the approval of the product should not be contingent on its potential impact on improvement and evaluation programs (Dekkers et al. 1995, i–ii).

The report on U.S. consumer reactions noted that there had been little change in consumption in the U.S. but that this had been achieved by not differentiating between treated and untreated milk. In states where consumers had reacted to the drug, rbST-free milk was marketed, due to consumer demand, but that demand now seemed to be declining. The exceptions to this were in Wisconsin and Vermont, where a dual-marketing system had been important in maintaining consumption levels. This seemed to have "started as much from farm and rural living concerns as from consumers" (Brinkman 1995, i–ii).

The Council of Canadians, an Ottawa-based public interest organization that had campaigned against Canada' s participation in the FTA and NAFTA and for the preservation of Canada's social programs, criticised the task force report and the government's conduct of the rbGH review. It questioned the task force trip to New York state, in which the scientist and the veterinarian interviewed by the group, Dr Dale

Bauman and Dr Les de Groff, had both worked closely with Monsanto. It also referred to recent scientific evidence on both animal and human health problems, including recent evidence on the bioactivity of IGF-I (Council of Canadians 1995a, 2).

On 18 June the House of Commons Committee on Health called for a two-year moratorium if rbGH was approved (McLaughlin 1995). In response, Health Canada re-stated its criteria for licensing animal drugs and reiterated its examination of health and labelling issues raised by the task force report. Health Canada also released its summary of the human safety evaluation of rbST, a review of the scientific literature on rbGH and IGF-I.

On 3 August the coordinator of the Toronto Food Policy Council, Rod McCrae, commented to the board of health on Health Canada's summary of the human safety data (which had been released in response to the call for a two-year moratorium). The major concern expressed was that it had ignored recent evidence disputing accepted knowledge, since most of the literature cited had been published before 1991 and therefore more recent work on the effects of the hormone had not been taken into account. The letter also stated that adoption of rbGH could prove hazardous to human health if people reduced their milk consumption in order to avoid problems associated with it; Health Canada had not considered this possibility in its review (Tabuns 1995). The board of health expressed these concerns to the Standing Committee on Agriculture.

The manufacturers were alarmed by these developments. In a letter to MPs dated 16 June (and marked "confidential") Monsanto Canada's vice-president of legal and public affairs, Ray Mowling, wrote that "I believe this issue is being used by the Council of Canadians and other special interest groups to dismantle the regulatory system" (Mowling 1995). Anticipating that a legislated moratorium would be imposed when the voluntary delay expired on 1 July Monsanto and Eli Lilly threatened to withdraw research investments from Canada if such a ban were implemented. Spokesmen for the companies said that Monsanto would consider reducing its expenditure of $6.5 million in Canada if a two-year moratorium were enforced and that Eli Lilly might shift funding from its Canadian subsidiary (Saunders 1995b, B7). The companies later denied having made such statements. They argued that the introduction of rbGH would be essential if Canadian dairies were to compete with increased imports from the United States as the provisions of the GATT and NAFTA gradually led to the erosion of Canadian import protection. A representative from Eli Lilly claimed that an inefficient regulatory system, open to pressure from external interests, created a less hospitable environment for investment in Canada (McLaughlin 1995).

Allegations had previously been made that Monsanto had offered a bribe to Health Canada to fast-track approval of rbGH. The CBC broadcast "Big Milk, Big Money, Big Muscle" on the *Fifth Estate* on 29 November 1994 alleged that four or five years earlier Monsanto had offered money to the department for "animal-related biotechnology research in Canada." A memo written by Dr Margaret Haydon, an official with the BVD, said that money was offered on condition "that the company receive approval to market their drug in Canada without being required to submit data from any further studies or trials." In December, Monsanto said that it had "proposed a commitment to future research based on potential sales of bovine somatotropin following its approval for use in Canada," but denied that it had offered a bribe to the department and demanded a retraction from the *Fifth Estate* (Bueckert 1994).

Soon after allegations of the bribe were broadcast, the health minister, Diane Marleau, faced conflict-of-interest questions about the director of the BVD, Dr Leonard Ritter. Dr Ritter had been on unpaid leave from the bureau since June 1993. As a witness at the standing committee hearings in March, he had said that he was a former public servant. On 7 December the minister said that she had asked the deputy minister to determine whether Dr Ritter had contravened conflict-of-interest guidelines (Ha 1994, A5). The Canadian Animal Health Institute later complained to the Ontario Press Council that a *Globe and Mail* article in which Dr Ritter's testimony had been discussed misrepresented the institute, but this claim was dismissed by the council ("Press Council" 1996, A11).

In August over fifty public interest groups composed a joint letter urging the prime minister to legislate a one-year moratorium in the event that Health Canada issued an NOC for Posilac (CIELAP 1994). The National Farmers' Union had also sent a letter to the prime minister in June expressing concern and trepidation over letters from Monsanto and Provel that "would seem to suggest alternatives for the circumvention of the Standing Committee on Agriculture" (Macklin 1994). As a part of its Safe Milk Campaign the Council of Canadians had encouraged members of the public to write to their members of parliament, the minister of health, and the executive director of the Dairy Farmers of Canada, expressing their opposition to the drug and suggesting that the federal government follow Europe's example and ban it until at least the year 2000.

In 1995 the Bureau of Veterinary Drugs declined Monsanto's application for the licensing of Nutrilac, due to indications of serious animal health problems and the inadequacy of the experimental design. It was noted that there had been a failure to show an increase in milk production when off-label medication was not used (rbST Internal Review Team 1998, 13). This rejection was followed in 1996 by a meeting between Health Canada officials and representatives from Monsanto and its Canadian subsidiary. At this meeting, it was agreed that a review of the data would be completed in the next year and that one principal reviewer would work full-time on the file, with the assistance from a team within the Bureau of Veterinary Drugs. It was later decided that a decision would be delayed until after a review of the human safety issues was completed in February 1998 by the Joint Expert Committee on Food Additives (JECFA), a committee established by the World Health Organization and the Food and Agriculture Organization (Senate Committee 1999a, 3). JECFA provides advice to the Codex Alimentarius Commission (known as Codex, or the CAC), which establishes food safety standards. In the case of veterinary drugs, Codex may establish a maximum residue limit that determines how much of a particular drug residue may be permitted in food. In the case of rbGH, no limit was specified by JECFA because it was concluded that the substance did not present a threat to human health. However, in both 1995 and 1997 a majority of countries voted to postpone the adoption of a Codex standard for rbGH based on JECFA's advice and pending the reevaluation of the scientific evidence (Comisa 1997) (for a further discussion of the Codex decision, please see chapter 5).

Although an NOC had not been issued by Health Canada, the Canadian Bureau of Veterinary Drugs had decided in 1990 that the human safety data were acceptable and that milk from treated cows would be safe for human consumption. The chief of the Human Safety Division, Dr Man-Sen Yong, had stated that the drug did not represent a risk to human safety, and Canada had voted accordingly at Codex meetings on the human health issue. Some of the evaluators in the bureau disagreed with this conclusion. One of them, Dr Shiv Chopra, told the director-general of the Health Protection Branch, Dr George Patterson, that if either he or Dr Yong made statements about human safety on behalf of the department, Dr Chopra and other scientists would "publicly oppose" them (Senate Committee 1998a). When Chopra asked Patterson to appoint one or two people to look into the rbGH evaluation, he agreed. Initially, Dr Chopra was appointed with Dr Gerard Lambert to coordinate this exercise, along with the rest of the Human Safety Division. Dr Patterson later expanded the team to include Dr Thea Mueller of the Therapeutic Products [human drugs] Division, and Mark Feeley of the Chemical Health Hazard Assessment Division. These six scientists comprised the Internal Review Team, and they were mandated to review the scientific data on human safety and to determine whether any "gaps" existed in the data regarding the risk to public health posed by the introduction of Monsanto's product. Patterson referred to the internal committee review as an "internal gap analysis" and noted that the gaps analysis report was reviewed by an rbST advisory team, whose comments were to be incorporated into the original report (Senate of Canada 1998).

During the internal review process, Chopra and five other scientists from the bureau filed a grievance against the department, which was eventually heard by the Public Service Staff Review Board (Eggertson 1998c, A6). The scientists claimed that they had been "pressured and coerced to approve drugs of questionable safety, including rbST" (see statement by Chopra, Senate Committee 1998a).[11] Chopra and other parties to the grievance offered to withdraw from the Internal Review Team, but Patterson advised them to continue in their role.

The conflict within the Bureau of Veterinary Drugs also created a split in the Internal Review Team, and two versions of the team's gaps analysis document were produced: one on 21 April 1998, the other in June 1998. The two members who were not from the bureau were concerned that the scientific issues had become entwined with the internal conflict and that the first report had placed undue emphasis on the internal conflict (Senate 1998b, Statement by Thea Mueller).

However, the six scientists did agree that their investigations had uncovered deficiences in the review process that warranted further consideration. Their major concern was that the standard toxicity testing had not been required in the rbGH case and that the reasons for waiving these requirements had not been adequately explained by the human safety evaluators. The Internal Review Team's critique hinged on a ninety-day rat-feeding study in which some rats had developed antibodies to the drug, suggesting the possibility that, contrary to expectations that the product would not survive digestion and enter the bloodstream, it had been orally absorbed. This study, the team claimed, contradicted the assumption on which conclusions about human safety had been based, thereby throwing the conclusions themselves into question. Although acknowledging that the indications were seen only at high doses, they believed that they warranted further consideration. The team were also concerned that regulators had not examined whether other hormones, growth factors, or residues could enter the milk supply as a result of drug treatment, nor had they thoroughly examined the human health implications of animal illness. These issues will be explored further in chapter 5.

Concurrent with the workings of the Internal Review Team, the Bureau of Veterinary Drugs had requested an external review process. The bureau asked the Royal College of Physicians and Surgeons of Canada and the Canadian Veterinary Medical Association to appoint two external panels on human safety and animal safety, respectively. Both panels' conclusions bore out the bureau's decisions up to that point. The Expert Panel on Human Safety agreed that the product did not represent a risk to human health. Although one of the members of the human safety panel, Dr Michael Pollak, commended the Internal Review Team for raising questions about the data, neither he nor the rest of the panel felt that further testing was warranted. The panel advised that the ninety-day study should be repeated and the results discussed with Monsanto scientists, but they did not feel that the evidence justified preventing the sale of rbGH in Canada on human safety grounds.

The Expert Panel on Animal Safety agreed with the bureau's conclusions on animal health and recommended that the drug should not be licensed in Canada. After these reports were presented, Health Canada announced that it would not approve the bovine growth hormone (Health Canada 1999).

In June 1998 the Senate Committee on Agriculture and Forestry began hearings into the rbGH review process and the allegations surrounding it. These hearings were primarily the result of action by Eugene Senator Whelan, a former minister of agriculture. The Senate hearings provided an opportunity for the scientists' grievances to be heard. There were allegations of management pressure, document shredding, theft of files, and witholding of the rbGH files. Although the hearings were called to investigate the rbGH evaluation process, the strongest evidence for management pressure was offered in the case of the beef steroid hormone, Revalor-H. One of the evaluators, Dr Margaret Haydon, had concerns about the human health implications of the test animal data. Neither she nor another evaluator nor her two supervisors (including Dr Chopra) recommended that drug for approval; but their views were overruled (Senate Committee 1999c).[12] Scientists' allegations about the animal drug review process have continued. In July 2001, Chopra and Haydon again told the media that their advice on animal drugs was being overruled by the head of the bureau, and they were again supported by four other scientists (Blueckert 2001, A4).[13]

The allegations of coercion to approve the drug could not apply in the rbGH case, because the grievors had either not been involved in the review or had been removed from it. Chopra reported that "I was kept from taking any direct part in the actual review. So incidentally were all of my colleagues in the human safety division" (Senate of Canada 1998b, Statement by Chopra). Although their removal was arguably a form of coercion, at the time the grievances were lodged they had been excluded from the process. Haydon had been removed from the rbGH review after an incident in which she had alleged that files regarding the evaluation had been stolen from her office. The RCMP had investigated the incident. The outcome of the investigation was not made clear in the rbGH hearings. Dr Margaret Haydon said that a report had been written by the acting director in the security services of Health Canada but that she (Haydon) had not been interviewed by the author of the report. She had reviewed rbGH files from three different companies between 1984 and 1994 (Senate of Canada 1998b, Statement by Margaret Haydon).

The allegations of coercion were rejected by the Bureau of Veterinary Drugs and Health Canada management, who pointed out that the Public Service Staff Review Board (PSSRB) had not verified them. The PSSRB concluded that the evidence presented by the scientists did not substantiate their claims of pressure and coercion by management, adding that it had no authority to examine the scientific issues arising from the hearings (McIlroy 1998, A5). However, it did note that "The evidence does show the presence of troubling scientific and interpersonal conflict in the BVD workplace" (Federal Court of Canada 2000, 15). At the hearings in October 1998, the recently appointed deputy minister, David Dodge, repeatedly acknowledged the problems within the Bureau of Veterinary Drugs ("there is a clear recognition that not everything is right") and his concerns about allegations of coercion. However, at a Senate committee hearing in 1999, he also pointed out that the information commissioner had investigated and dismissed allegations of document shredding and that a review had found no evidence of pressure by managers to approve veterinary drugs. In relation to the witholding of the rbST files from scientists in the department, he said tightened security had been introduced in response to the reports of stolen files. Dodge stated that the "sole job" of the department was to protect the health and safety of Canadians. However, he also added that regulators were subject to pressures from industry and to a work load that had risen without a corresponding increase in the number of scientists to handle it.

The Senate committee itself was unable to comment on the accuracy of the scientists' allegations. However, it recommended that an investigation of the drug approval process be carried out; Dodge pledged that the Science Advisory Board would undertake this exercise.

Dodge emphasized that efforts were being made to address these issues through the Health Protection Branch transition process begun one year earlier. The transition process was intended to strengthen the science underlying regulatory decision-making, improve and modernize the country's health surveillance capacity, improve risk management "while recognizing the roles and responsibilities of partners and participants," update and integrate federal health protection legislation, and improve the delivery of health protection programs (Health Canada 1998a, 3). The review of legislation involved the replacement of existing pieces of relevant legislation, now numbering more than twelve, with a single, general statute tentatively entitled the Health Protection Act, which was intended to be tabled in the year 2000 (Health Canada 1998b, 2). A committee was also developing a "risk management framework," to provide a basis for departmental policies.

A restructuring of the Health Protection Branch was regarded as necessary because of the emergence of new health risks, the existence of new technologies, including communications technologies that can transmit knowledge rapidly, and the distribution of work formerly done by Health Canada to departments and organizations outside it. The creation of the Canadian Food Inspection Agency – a body established in 1997 to enforce food and safety standards created by Health Canada – was also cited as requiring the rethinking and restructuring of Health Canada's role. With regard to risk management, the development of a framework was regarded as necessary in order to define the responsibilities of multiple partners, to consider a number of determinants of health and focus on the most significant of these, and "to respond to risk in a holistic way, considering not only the risks associated with individual diseases, agents, or products in isolation, but also their relative importance in the context of other risks. This will help to set priorities for allocating limited resources" (Health Canada 1998c, 1).

The transition process is not yet complete; however, it has been criticized by the Canadian Health Coalition (CHC), a public advocacy group dedicated to the preservation of a public health care system. The CHC argues that the transition is an attempt by the federal government to abdicate its responsibility for health care by distributing it among the other bodies that have a role in health protection and surveillance, such as the provinces and the universities. The CHC identifies public protection from food and drug hazards as a basic component of public health care and criticizes the shift toward the language of risk management, rather than public protection, as undermining public health by encouraging regulators to view risks in economic terms, thereby comparing the risks involved in approving a product to the potential benefits that may accrue from it (Canadian Health Coalition 1998). At the hearings, David Dodge had stated that the budget of the Health Protection Branch would be increased by 70 percent between 1997 and 2002. In 1999, $65 million was restored to the branch over three years for food safety research (Health Canada 1999b, 2). However, it was not until 1997–98 that funding had been (almost) restored to the 1993–94 level of $237 million. During the period of cutbacks, critical in-house research capacity had been lost (Canadian Healthcare Association 1998, 2): the entire Bureau of Drug Research had been eliminated in 1997, and many of its scientists were lost to the United States (Canadian Health Coalition 1998, 4). Wiktorowicz has argued that the changes implemented through the Health Protection Branch transition process may not adequately protect the public interest:

> From a public-health perspective ... the criteria used to develop organizational and policy change at the HPB have provided a sub-optimal basis for reform, due primarily to the many forms of market failure to which the regulation of pharmaceuticals is subject. For example, the *partnership* and *efficiency* criteria that guided policy renewal have led to the transfer of important responsibilities to partners, with the potential for either a conflict of interest or inadequate information, for which the legal basis is not always clear. The resulting realignment of the HPB's roles and responsibilities may be characterized as leading to a shift from a comprehensive approach to public-health protection to one based on strategic risk management, with responsibilities dispersed among government, industry, academia and consumers. The rebalancing of goals in the redesign of the regulatory process suggests a change in the role of the state in the context of public-health protection and highlights issues of concern to the public interest that may not be fully recognized as deregulation occurs in other sectors of the economy. (2000, 1)

The transition had not been in process for long enough to become the focus of the Senate hearings. However, the issue of Health Canada's relationship to industry, of concern in the transition process, came under scrutiny. Bureau managers explained that since cost recovery had been introduced to the bureau in 1996, 70 percent of the cost of drug review had been borne by the companies. (The Monsanto rbGH application had been exempted from this requirement, since it had been lodged before the policy change). Before the introduction of the cost recovery program, the industry and the bureau had agreed that a joint management advisory committee would meet regularly "to share and discuss issues of common interest and importance to the veterinary drugs program." The committee was comprised of the chiefs of various divisions within the bureau and representatives from the industry body CAHI, and it regularly included the drug evaluators. According to the Senate committee, it was at these meetings that the names of drug evaluators had been given to the sponsoring companies (Senate Committee 1999a). In the same year, the privacy commissioner found that Health Canada had violated the confidentiality rights of a former Health Protection Branch scientist, Dr Michele Brill-Edwards, by disclosing her personal records to the assistant deputy minister (Ditchburn 1998, A4).

The committee expressed its concerns about the Joint Program Management Advisory Committee in an interim report produced in March. It urged the minister of health to review the composition and role of the advisory committee. It also recommended that the Minister initiate a review of the drug evaluation process in the department, either in conjunction with the auditor-general or subject to his or her review. The committee noted that it had every confidence that the government would carry out the necessary measures (Senate Committee 1999a).

The revelations from the rbGH case highlight the difficulties in the science-policy relationship and the multiple pressures to which regulatory scientists are subjected. In this case, however, the issue cannot be reduced to a simple conflict between "science," on one hand, and "industry interests," on the other – for two reasons that will be explored further in the next chapter. One reason is that on the recommendation of its scientists, who questioned the legitimacy of the data and the experimental design, the department consistently rejected the drug on animal health grounds. The second reason is that in Canada and the United States, both regulators and expert advisory committees have agreed that there is little likelihood of risk to human health. It seems unlikely that the human health questions would have been resolved by the submission of further scientific evidence.

What the case does indicate, however, is that given existing pressures, the regulatory system does not readily engage in questioning prevailing scientific knowledge or in creating broader conceptions of health that, in the case of rbGH, could link human and animal health. The interpretation of the rbGH data created difficulties within the bureau even though it was not subject to cost recovery and even though evaluation had begun before the funding cuts after 1993. The institution of cost recovery, the cessation of in-house research funding, and the creation of the Joint Management Advisory Committee do not bode well for a regulatory science that is more open to unorthodox questions and more responsive to public concerns. It is already difficult, given scientists' time constraints and understanding of corporate requirements, to question the assumptions of existing knowledge and to raise concerns that might seem unreasonable to some. It is difficult to see how the industry share of the multiple pressures these scientists are subject to cannot be increased when industry is responsible not only for the creation of the material under review but also for the funding for its evaluation.

CONCLUSION

Both the Canadian and the U.S. regulatory systems exist in the context of the broader political economy. In the United States, changes in the political economy – particularly industrial restructuring around technological innovation – were facilitated by government policy and judicial decisions.

In Canada, the state has accepted that technological innovation is best facilitated in collaboration with the private sector and has developed strategies to foster the biotechnology industry, including the provision of financial support. It has also followed the U.S. example in regulating the products of biotechnology through existing statutes rather than a separate statute, although products not covered by existing acts will be covered under the CEPA. However, until the year 2000, Canadian courts refused to recognize the patentability of higher life forms, and other aspects of patent law are not in conformity with Canada's trading partners. Although the CEPA was not amended in accordance with critics' recommendations, the process of reviewing the act generated debate about the future of biotechnology regulation. Action has so far been limited, but government advisory committees have recognized the importance of public input on the future direction of biotechnology policy.

With respect to the Canadian dairy system, GATT and NAFTA have not yet had a direct impact, since quota restrictions have been replaced by tariffication. However, some of the concern about rbGH in Canada has been regarding the extent to which the drug would affect structural adjustment, which will take place as subsidies are phased out.

There have been fewer mechanisms available in Canada than in the United States through which these kinds of concerns could be introduced into the regulatory process. However, the expression of reservations about the drug evaluation process by scientists within the Bureau of Veterinary Drugs drew attention to the evaluation of rbGH, and its investigation by the Senate allowed various groups to voice their concerns. Nevertheless, it is difficult to determine whether this degree of public scrutiny had an influence on the decision. Ultimately, however, the government followed the advice of the rbGH reviewers, on both animal and human health, after the BVD's conclusions were endorsed by an external panel. The evidence presented at the rbGH hearings, however, indicates that the evaluation process should be investigated further, because such advice has not been heeded with regard to other animal drugs.

5 The Scientific Debate

INTRODUCTION

Since the late 1980s, the public debate about rbGH has focused on safety issues. Would milk from treated cows be safe to drink? Would its consumption lead to an increased risk of breast or other cancers? If animals become sick when treated with rbGH, will people consuming their milk also be affected?

In spite of the ferocity with which these issues were argued, there was a surprising degree of consensus about the scientific issues involved.[1] It was agreed that the evidence showed little likelihood of risk to human health; however, it was also argued that the existing data were insufficient and that results that contradicted scientists' expectations should be investigated further, rather than explained within the framework of existing scientific knowledge. It was agreed that there was a statistically significant increase in animal health disorders. Proponents and regulators of the drug in the United States believed that the increase was small and that occurrences of illness could easily be managed by farmers. They also believed that the increase in the incidence of animal health problems did not pose any threat to consumers' health. Critics, on the other hand, argued that the incidence was not minor and that it was not manageable, a position that Canadian regulators agreed with. The company and university scientists argued that the risk was nonexistent when milk production was taken into account.

The debate therefore was not about the evidence so much as its meaning. What level of testing is adequate to draw conclusions about human and animal health? How should statistically significant results be interpreted? What level of risk to animal health is acceptable? In order to answer these questions and to determine what data were necessary in order to answer them, scientists made assumptions based on their knowledge of the scientific literature and of existing dairy practices. These assumptions determined whether they thought the data were sufficient to warrant approval of the drug.

The first of the following four sections analyzes the human health debate, examining the positions of the proponents of the drug and its critics. It also compares their interpretations with those of scientists outside the debate. The biomedical researchers interviewed for this study had little or no involvement with the debate when they carried out their investigations. However, their work was later used to support one side or another; sometimes both. The second section that follows analyzes the animal health debate, and the debate about the human health implications of the conclusions from this argument is the subject of the third section. The fourth section contrasts the scientific decision making in North America with that in Europe and the way the trans-Atlantic conflict spread into the international food-safety standard-setting body, Codex.

THE HUMAN HEALTH DEBATE

The human health debate focused on two issues. One was whether recombinant bGH was more likely than the natural bGH already present in milk to cause human health problems, particularly allergic reactions. The second issue, which caused the greatest public concern, was whether a hormone-like substance called insulin-like growth factor-I (IGF-I), which mediates the action of growth hormone and is elevated in the milk of treated cows, would present a health risk. Research conducted during the period of the rbGH debate – although conducted quite separately from it – provided evidence that insulin-like growth factors played a role in the development of some malignant tumours, including breast cancer. Epstein (1990a, b; 1996) raised the alarming possibility that the widespread consumption of milk from treated cows was a potential cancer risk. This suggestion and the publication of new work on IGF-I, intensified the controversy around the drug and led to demands for the reexamination of prior regulatory decisions. Another important issue was whether the widespread use of the drug would result in higher antibiotic usage and hence greater risk of consumption of antibiotic residues in milk, leading to resistance in the general population. Since this question is connected to animal health concerns, it will be discussed in the section on animal health.

Natural vs Recombinant Bovine Growth Hormone

Bovine growth hormone is composed of a sequence of 191 amino acids. The recombinant hormone developed by the companies had a slightly different amino acid sequence from the natural molecule.[2]

Recombinant rbGH was the first protein hormone produced for use in food animals. FDA scientists expected that it was extremely unlikely that rbGH would produce any effects in humans, because the natural bovine hormone had been ineffective in the treatment of human growth hormone deficiencies. Like other protein hormones, growth hormone binds to receptors on the cell surface, which then send chemical signals to the cell. *Bovine* growth hormone is sufficiently different from human growth hormone that it will not bind to human receptors. Its activity is species-limited; humans can respond only to growth hormones from other primates, whereas rats, for example, can respond to growth hormones from a number of species.

The reviewers' determinations about the data submitted by the companies and their interpretation of that data were influenced by their reading of the existing scientific literature. In order to determine whether rbGH residues presented a risk to human health, FDA scientists reviewed the reports from the 1950s on attempts to develop a biologically active form of bGH for human use. At that time, researchers had not been successful in their efforts to treat dwarfism and other growth hormone deficiencies with either the hormone in its natural sequence or with other combinations of its component amino acids. According to one of the FDA scientists, "the general feeling was there were no safety issues with bGH itself. A lot was known about it. So I don't think there was any doubt in anyone's mind that bGH was never going to be active, so it was never going to be any kind of safety issue."

FDA scientists also reviewed the literature on protein digestion and absorption in adults and newborns that indicated that in adults most proteins are not orally active; that is, they are broken into smaller peptides or amino acids during digestion and do not enter the bloodstream intact. When proteins are broken down into their constituent amino acids, they can no longer cause an adverse reaction in humans. The results for newborn infants were conflicting; however, where evidence of protein absorption existed, the amount absorbed was insignificant (Juskevich and Guyer 1990, 876).

Although agency scientists believed that rbGH would not be biologically active in humans and that it would not be absorbed, they requested several oral toxicity studies in order to determine whether this was correct. The longest of these studies was a ninety-day oral toxicity study by Monsanto that compared rats injected with rbGH to those receiving it orally. At the low dose of rbGH given orally, absolute spleen weight increased, but the FDA reviewers decided that this finding was incidental, because it was not dose-dependent; that is, the finding was not more pronounced as the dose increased. Increases in ratios of organ weight to body weight were sporadic and were also interpreted as incidental. It was therefore concluded that rbGH was not orally active (Juskevich and Guyer 1990, 878) and that milk from treated cows could safely be released into the food supply.

This ninety-day study by Monsanto included a discussion of another group of rats that had been investigated to determine whether rbGH might be absorbed through the gut. However, these data were not reviewed by the FDA when they were originally submitted (FDA 1999, 4n) but were made public by Health Canada scientists almost ten years later, after the controversy had been investigated by the Canadian Senate.

Health Canada's human health evaluation reached the same conclusion as the FDA. According to a former Health Canada scientist, when evaluating an animal drug, the primary concern is human safety: "If for any reason the Human Safety Division said no, this is a potential danger to humans, especially to children, it's not going to be marketed, that's it."

As in the United States, it would appear that normal scientific understandings about proteins influenced the human health decision, which was made by the chief of the Human Safety Division, Dr Man-Sen Yong, who determined in 1990 that the use of the drug did not represent a risk to human health. The former Health Canada scientist accepted Dr Yong's reasoning, and agreed that because the product was a protein hormone, "it had no potential for causing human health problems." The scientist was aware that concerns had been raised about a human health risk from increased levels of growth factor, but believed that there was no scientific justification for those concerns.

This decision was disputed *within* the Bureau of Veterinary Drugs, however. As discussed in chapter 4, an evaluator in the Human Safety Division of the BVD, Dr Shiv Chopra, raised questions about the safety of the drug and the adequacy of the review process. A standard data package had not been submitted for the human health evaluation of rbGH, because reviewers had concluded that short-term tests indicated that the product was safe. However, not all of the data had been reported properly, and there were anomalous results among the data.

The crux of the internal review team's criticism was data from the ninety-day oral toxicity study carried out by Monsanto, which showed that at extremely high doses, some rats developed antibodies to rbGH, indicating the possibility that rbGH *had* been orally absorbed. Between

15 and 20 percent of rats treated with 5,000 and 50,000 nanograms of rbGH per kilogram per day produced antibodies; in addition, one rat in the lowest dose group also developed antibodies, but Monsanto scientists suspected that this result was due to a mislabelled sample. Monsanto scientists explained that the gut may have produced the antibodies in response to the sheer quantity of the drug introduced or that the daily tubing of rats may have caused breaks in the gut lining and led to the direct entry of rbGH into the animal's bloodstream (rbST Internal Review Team 1998, appendix 6).

However, the members of the Canadian review group argued that any evidence of oral absorption, even at high doses, necessitated further consideration because it contradicted the expectations on which the decision was based.

The expert panel on the human safety of rbST appointed by the Royal College of Physicians and Surgeons of Canada disagreed; they did not believe that the results of the study meant that the drug warranted long-term human safety testing. They thought that the ninety-day study should be repeated and that the bureau scientists should discuss the results with the Monsanto investigators. If this effect occurred when the study was repeated, it would suggest the possibility of "occasional hypersensitivty reactions" to people drinking milk from rbGH-treated cows. However, there was "no biologically plausible reason for concern about human safety if rbST were to be approved for sale in Canada" (Expert Panel 1999, iii).

Health Canada's internal review committee was concerned about the lack of explanation for and discussion of the decisions that were made. The antibody results were not included in Monsanto's summary of the ninety-day study, and the results themselves were placed at the end of the data package, so that "we had to go through all the pages to find ... that study" (Senate 1998b, "Statement by Gerard Lambert"). The gaps analysis report noted that Monsanto's forty-volume submission on human safety "was reviewed by the Human Safety Division in less than two weeks" (rbST Internal Review Team 1998).

It is impossible to know exactly what transpired within the Human Safety Division review process and therefore why more detailed explanations were not written at the time of the review. However, it seems plausible that if the reviewers began with an understanding that rbGH was not biologically active in humans and was not likely to be orally absorbed, they would not have thought it necessary to elaborate on indications of absorption at very high doses.

The FDA did not offer an explanation of why the antibody data were not reviewed during the evaluation of the Monsanto submission. However, in a response to the criticisms of the FDA that followed the release of the Canadian gaps analysis report, the agency responded that the Canadian scientists had interpreted the results incorrectly. Although there was antibody evidence suggesting oral absorption, the study did not provide sufficient evidence to decide this definitively, nor did it provide any indication that rbGH had any other effects on the animals, since body weight and organ weight did not change with the dose of rbGH administered. The agency argued that because the concentration of rbGH that people, particularly children, would be exposed to was hundreds of times less than the concentration that caused the antibody reaction in rats, long-term studies were not necessary (FDA 1999, 2).

Insulin-Like Growth Factors: A Potential Cancer Risk?

When cows are treated with rbGH, there is an increase in the level of insulin-like growth factor I (IGF-I) in their milk. IGF-I is a protein that is normally present in body tissues and affects growth, development, and metabolism (Stewart and Rotwein 1996). In both cows and humans, IGF-I is produced by the liver in response to growth hormone stimulation. Like a hormone, it circulates in the blood to effect changes in cells distant from the liver. (When circulating in the blood, IGF-I is bound to carrier proteins.) However, unlike a hormone, IGF-I may act upon the same cell that produced it, or it can act on a nearby cell (Pollak et al. 1987, 223).

The effect of IGF-I has been of concern because, unlike growth hormone, bovine and human IGF-I are structurally identical (Sara and Hall 1990). The consequences of potentially elevated milk IGF-I levels has been an explosive public health issue because of Epstein's suggestion that this could lead to an increased risk of breast cancer. In the late 1980s medical researchers had begun to investigate the possibility that the insulin-like growth factors may play a role in the development of certain kinds of tumours; some research suggested that higher levels of IGF-I in circulation were associated with a higher incidence of particular types of cancer. This research has continued into the 1990s. However, there are many questions that have yet to be resolved about the relationship between growth factors and cancer. Scientists examining this question have also stressed that a relationship between IGF-I in circulation and tumour growth does not necessarily imply a relationship between the consumption of milk from treated cows and cancer. There were several points of debate regarding this relationship. The first point was whether levels of IGF-I were actually higher in the milk from treated cows. The FDA summary published in *Science* stated that the milk from treated cows contained up to 25 percent more IGF-I, an assertion that was later accepted by a National Institutes of Health review panel.

Monsanto, one the other hand, argued that more recent studies showed that there was no significant difference in IGF-I concentrations. In support of this claim, Monsanto cited a study by the World Health Organization/Food and Agriculture Organization Joint Expert Committee on Food Additives (JECFA) that stated that "the most definitive and comprehensive studies show that IGF-I concentrations are not altered after rbST treatment" (1993, 41, quoted in Collier et al. 1994). Some university scientists viewed the increase in comparison to the population of untreated animals and argued that since cows have ten to fifteen times the average level of IGF-I in their milk just after they have calved, an increase in the growth factor induced by rbGH would not be "meaningful." Critics, however, rejected this comparison. T. B. Mepham, professor of bioethics at the University of Nottingham, disputed the position that IGF-I increases fall within a range of normal variation, noting that widespread adoption of the drug would lead to a mean increase in growth factor concentration. A review article by several Canadian animal scientists has also claimed that the majority of studies showed an increase in IGF-I concentration (Burton et al. 1994, 189).

The next point of debate was whether an increased level of the growth factor in milk was likely to have any impact on humans. The FDA had not examined the IGF-I question before authorizing a zero withdrawal period for milk and meat from treated cows. However, when the safety of ingested IGF-I became an issue, a scientist in the Toxicology Branch undertook an extensive literature search on all growth factors that could possibly be affected by growth hormone and decided that "the only one that was of any concern was IGF-I." Basic research indicated that it was difficult to obtain an effect by injecting the growth factor, because binding proteins "mopped up" additional circulating IGF-I, rendering it inactive. The scientist also searched the literature on protein absorption in adults, newborns, and children, as well as the literature on the oral activity of insulin, which has a similar protein structure to IGF-I: "The picture that came from the whole thing from my point of view was the likelihood that you were going to see any significant safety problems from IGF-I was nonexistent. It was a protein; there was no indication that it would be absorbed or that it would not be broken down after you take it orally; and even if some minor amount got absorbed, it wasn't enough to produce an effect. Plus there are already levels of growth factor in milk, and therefore any increase would probably be negligible in terms of the residues."

The paradigm about the biological activity of bGH was extended to IGF-I, about which much less was known. As one scientist put it, "there weren't 30 years of studies about IGF-I"; however, since it was a protein, data about IGF-I could be assimilated into the paradigm about protein absorption, and the studies were therefore the same length as most of the rbGH studies. The scientist was so convinced about the unlikelihood of health problems from ingested IGF-I that she felt oral activity studies were unnecessary. However, the director of the FDA's Center for Veterinary Medicine advocated laboratory tests for IGF-I.[3] In retrospect, it was felt that this decision was correct, not for scientific but for political reasons: "It was apparent that the scientific reasoning behind what we did really didn't matter to people. No one seemed to be listening much to the science to start with, so I think it was a very wise decision that those studies were done. They showed us exactly what we expected; it wasn't orally active, well we didn't expect it to be orally active – and at what levels."

At the request of the FDA, Elanco and Monsanto conducted two-week oral toxicity studies with recombinant IGF-I in rats.[4] When proteins are given therapeutically, the effects are visible within a few days; so a study of fourteen days was regarded as sufficient to establish the oral activity of IGF-I. At the end of the treatment period, the rats' body and organ weights and their bone dimensions were measured. On the basis of this study, it was determined that rIGF-I is not orally active at doses up to 2 mg/kg per day (Juskevich and Guyer 1990, 881).

Some clinical signs were seen in orally treated rats, but these were not interpreted as treatment-related. One block of male rats treated at the highest dosage showed a slight but significant increase in body weight in the latter half of the study, with a significant daily weight gain (880). In an interview, a former Monsanto scientist pointed out that this sign occurred only in the high-dose group that had started treatment on day two; the high-dose group that had been initiated on the first day did not have a significant weight increase. Male rats in the other dosage blocks did not show weight increases, nor did any of the females. Nor were serum levels of IGF-I increased in the high-dose group.[5] It was therefore concluded that this sign was probably not treatment-related (Juskevich and Guyer 1990, 880–1).

These statistically significant counterinstances and several others mentioned in the *Science* article were not surprising. According to one scientist, "if you do enough measurements, in any toxicology study that you look at, something pops up as statistically significant." What the reviewer had to decide upon was whether a statistically significant result was also biologically significant, that is, whether it could be attributed to the drug.[6] In order to be regarded as biologically significant, the effect needed to fit several expectations. First, if an effect was apparent at a lower dose, it would also be expected at a higher dose; indications observed only at a lower dose level were therefore regarded as insignificant. Second, if the effect occurred in only one group of animals and not in another, the effect was more likely to be assigned as a random effect. Third, since the drug has various related effects, if a change was exhibited in only one parameter without concomitant changes in other parameters, it was not attributed to the drug. The reviewer's knowledge of the existing literature enabled him or her to differentiate between facts that were biologically significant and those that were insignificant; from within the paradigm, anomalies could remain anomalous, and further explanation was not necessary. The scientist was aware that this method was problematic for the critics and the public, "because it's one of those things that almost seems like scientific magic, where you make things go away that you don't want to see, but it isn't; it's a time-worn method of going through these studies and determining what's statistics and what's real."

The critics cited the counterinstances in Juskevich and Guyer's report as evidence that, contrary to the authors' conclusions, IGF-I was orally active. The British critic T.B. Mepham questioned their dismissal of these results, arguing that the counterinstances should be taken at face value or the experiments should be repeated. Mepham calculated that if a 10-kilogram infant consuming one litre of milk per day were exposed to the highest concentration of milk IGF-I reported – 25 µg/litre – this would represent one-eighth of the dose shown to effect changes in clinical parameters in rats. However, since a safety margin of 100 times the dose effective in animals should be applied to human subjects, Mepham argued that infants would be exposed to 12.5 times the recommended minimum (Mepham 1992, 737).

Epstein repeated these claims in support of his argument that infants would be exposed to increased health risks. He posited that the introduction of rbGH would potentially lead to "premature growth stimulation in infants, gynecomastia in young children, and breast cancer in women" (1990, 80). He drew on the work of cancer scientists examining the role of IGF-I in the development of various kinds of cancer, including breast cancer and advanced this position on the grounds that "There is unequivocal evidence that a wide range of intact proteins are absorbed across the gut wall in a wide range of species including humans" (1996, 175).

The third point of debate was about the *amounts* of IGF-I that could potentially enter into circulation or come in contact with the gut lining. One university research scientist objected to Epstein's statements because he believed that the concentration of IGF-I in milk from rbGH-treated cows was too low to be biologically active in humans and because the level of IGF-I in circulation implies very little about the growth factor's local activity: "To talk about the concentration of IGF in blood or milk is really silly because it has little relevance at the tissue site of action." He had researched the relationship between insulin-like growth factors and breast cancer. However, in his view, the complexity of the system and the amounts of IGF involved made it extremely unlikely that IGF-I concentrations in milk would have any adverse effects on human health. "That's the main argument from the scientific point of view about the whole IGF furor, and people like Rifkin et al. are just blowing it up without any scientific background. It's fear-mongering. My works linking IGF to breast cancer then get transposed to BST hysteria and nobody's stopped to think, what are we talking about here? We're talking about small amounts of a very potent peptide, certainly, but the concentrations are so small, and not any different in the treated or untreated state, that it wouldn't have any major impact in my opinion."

FDA scientists and the Canadian external review panel looked at the quantities of IGF-I that were likely to be absorbed in relation to the amount of IGF-I already present in the body. One FDA scientist commented that "For instance, there's a lot of free IGF in saliva, so if you believe that IGF is absorbed every time you swallow you increase your plasma levels of IGF. So, obviously then, it's not a problem, because if it were, you'd have all kinds of growth effects from doing that. This is not proof; but this is an indication that it's not a big deal. So when you look at quantities in saliva compared to milk, it's probably not going to be important." The FDA also looked at whether IGF-I would be present in infant formula and compared its levels in cows' milk to those in human milk.

From the scientist's perspective, the critics' argument that increased IGF-I levels in the milk of rbGH-treated cows could present a human health risk was nonsensical: "There are postulations that make absolutely no sense based on what's known in particular areas of science. I thought the arguments they were raising were just out there somewhere."

The FDA attempted to deal with public concern and requests from the dairy industry for a third-party review by asking the National Institutes of Health to convene a technology assessment conference. The NIH review reassured many of the critics who felt that their questions about the absorption of IGF-I had been addressed by the panel (see *Animal Pharm* 1990, 12; Mepham et al. 1994, 1446). By 1994 most critics were no longer concerned that rbGH treatment would lead to an increased risk of breast cancer. However, there was still uncertainty regarding whether IGF-I in milk could have effects on cells in the gut that came into immediate contact with it. Although the NIH had not regarded the approval of the drug as a human health risk, it admitted the possibility that IGF-I in milk could affect the gut lining: "The importance of the increased amounts of IGF-I in milk from rBST-treated cows is uncertain ... Whether the small additional amount of IGF-I in milk from rBST-treated cows has a local effect on the esophagus, stomach, or intestine is unknown" (NIH 1991b, 1425).

In the meantime, recent studies established evidence that IGF-I could have an effect on the cells lining the gut and that proteins present in milk could protect growth factors from digestion. A study by Olanrewaju et al. (1992) established evidence that the cells lining the gastrointestinal tract grew when exposed to IGF-I at levels equivalent to those in bovine milk (Epstein 1996, 180). Another study produced evidence that casein – a protein found in milk – protected a different growth factor, epidermal growth factor, from digestion; this led to the inference that IGF-I in milk would be more likely to survive digestion than the growth factor alone. On the basis of this evidence and the NIH report Epstein suggested that rbGH-milk consumption could affect the risk of gastrointestinal cancer.

In 1995 a study by Xian et al. posited that casein may protect IGF-I itself from degrading during digestion. This study provided additional support for the inferences made from the studies on epidermal growth factor. The Toronto Food Policy Council drew on this finding in its campaign against the introduction of rbGH (Tabuns 1995, 1). In an interview, however, Xian has stressed the difficulty of drawing conclusions for human health on the basis of his studies, emphasizing that they may not have any relevance outside the laboratory. Another study that showed that casein protected IGF-I was published in 1997 (Scientific Committee on Veterinary Measures Relating to Public Health, 17).

With regard to the impact of IGF-I on the gut – and other questions related to IGF-I – the NIH and the Canadian Expert Review Panel implicitly recognized a distinction between regulatory and research science. Although the NIH panel advised that further research should be conducted to "determine the acute and chronic local actions, if any, in the upper gastrointestinal tract" (NIH 1991b, 1425), it also stated that "it did not consider that decisions on the commercial use of rBST should be delayed until these studies are completed" (NIH 1991a, 231). This would imply that the NIH recognized the limitations of the FDA's mandate and that it recognized a distinction between "regulatory" and research science.

From the regulatory scientist's perspective, there are always more questions that could be researched; her job is to make a decision from the information that she has available. The scope of the studies and the context in which they are interpreted is therefore determined by existing knowledge: "You can only design studies that you think are the most relevant to determining whether you're going to have any problems with something, and take the available information from the literature to get a picture of things ... And there will always be people who will say it's that one extra molecule that's going to put you over the edge, and you can always make that argument, and I can never say well you're wrong, but based on logic it doesn't really make any sense."

The reviewers responsible for the human safety decision at Health Canada perceived the issue in a similar way to their American counterparts. However, the Internal Review Committee raised questions about IGF-I. As with rbGH itself, the scientists were concerned that the absence of the standard data package meant that the oral absorption of IGF-I had not been sufficiently investigated. At the Senate committee hearings on the Health Canada review process, Dr Thea Mueller argued that the level of IGF-I that consumers would be exposed to needed to be more accurately determined before deciding on what kinds of toxicological studies needed to be performed. Are the reported levels of IGF-I an accurate representation, or could the levels be higher than currently reported? What happened to growth factor levels in cows over a prolonged period of time? Would these levels rise significantly? Were the methods used to measure it sufficiently similar that results from different studies could be compared? How do milk processing practices affect it? Dr Mark Feeley acknowledged that if all the IGF-I in the milk from treated cows was absorbed, it would amount to less than 0.1 percent of the growth factor already in circulation. However, he thought that although this amounted to only a fraction of the total amount, it was not clear whether it would have an effect on a particular tissue that was exposed to it when it was absorbed (Senate Committee on Agriculture and Forestry 1998b).

On the other hand, scientists who had concluded that the risk to human health was so low that it did not warrant regulatory action argued that "absolute safety is not guaranteed" and that the safety margin in the rbGH case was "wide enough to warrant putting resources elsewhere." When compared with the amounts of IGF-I already present in body tissue, the increase in IGF-I in milk amounted to "a drop in the ocean": "2000 nanograms of IGF-I in a sea of 10 million nanograms per day" (see the statement by Dr Stuart Maclean, Senate Committee on Agriculture and Forestry 1998c).

At the Canadian Senate hearings, it became apparent that the differences in opinion were about the appropriateness of particular kinds of regulatory action. Like the NIH panel, the members of the Expert Panel on Human Safety advocated more general research on the relationship between IGF-I and cancer. As scientists, one panellist pointed out, they supported further exploration of scientific puzzles. One of the members of the expert panel, Dr Michael Pollak, conducts research on the relationship between IGF-I and breast cancer. Not long before the Senate hearings, Pollak had published evidence that men with high IGF-I levels have a greater risk of prostate cancer ("Health Monitor" 1998a, 58; Chan et al. 1998). However, Pollak argued that although high levels of circulating IGF-I were related to cancer, this did not necessarily mean that drinking milk from treated cows would pose a risk, because it would not affect IGF-I levels in circulation. The regulatory question was different from a basic research question, and in his role on the expert panel Pollak was examining the validity of the regulatory decision. The panel was confronted with the problem of whether the evidence on human health could "justify the Canadian government in keeping BST out of Canada," and, they believed, it did not. The chair of the expert panel, Dr Stuart McLeod, said that "we did look at the Gaps Analysis and decided that most of those gaps were unimportant" (Senate Committee on Agriculture and Foresty 1999a).

The panel noted that the complexities of the IGF-I issue had been simplified in the past and pointed out that absorption of IGF-I had occurred in some experiments and that casein resulted in greater likelihood of absorption (results that had been discovered by critics in the mid-1990s). In the view of the panel, however, proof of the absolute safety of milk from treated cows could be obtained only from follow-up studies on a population that had been consuming it. To demand data from such studies *before* approval "would represent a standard of certainty of safety which does not currently exist for any other food product" (Expert Panel on Human Safety of rBST 1999, 16). The impossibility of knowing the impact of rbGH until it had been released was what concerned Dr Shiv Chopra, however. Chopra said that "when you use something like BST in cows for economic purposes, there is no way to know now what it will do." Although Chopra conceded that "adverse effects are not anticipated," he argued that "actual proof was not provided" (Senate Committee on Agriculture and Forestry 1998b).

Perceptions of the acceptable level of risk depended upon whether the individuals thought that a higher standard needed to be adhered to for a product that did not offer any benefits to the consumer. Members of the Health Canada Internal Review committee and the senators hearing their testimony, frequently pointed out that the purpose of rbGH was economic, not therapeutic; under these circumstances, the greatest caution should be taken. Whereas the Canadian Expert Panel on Human Health stated that given current methods of analysis, "it is impossible to prove danger" and that the requirement of absolute safety would represent an unfair standard for bovine growth hormone and other similar biotechnology products (Senate Committee on Agriculture and Forestry, 1999a: 53), the members of the internal review committee wanted assurances of absolute safety before proceeding, *given the nature of the product*. Mueller said that "we should leave no stone unturned in this instance because we are dealing with a basic food stuff such as milk" and argued that once exposure levels had been determined, the analysis should take into consideration the benefit to the consumer: "if the consumer benefit is small, then the risk should be actually negligible or zero because why should they be exposed to any kind of risk if there is no benefit?"

As with rbGH itself, the scientists' concerns about IGF-I stemmed from the lack of detailed discussion of the rationale for the original bureau decisions on human safety. Dr Gerard Lambert of the internal review team told the Senate committee that "the IGF-I study was not reported properly." The study of 1990, which showed some evidence of oral absorption of IGF-I, "was not described in totality" (Senate Committee on Agriculture and Forestry 1999b). Dr Michael Pollak (of the expert review panel) agreed that the rbGH case was "not a rubber stamp issue," and commended the internal review team scientists for urging further thought. However, having had the opportunity to examine the data further, the expert panel did not believe it was necessary to carry out the team's recommendation of further testing.

In contrast to regulatory scientists, researchers engaged in basic or medical research do not operate under the same pressures. The scientists engaged in basic and applied medical research interviewed for this study proceeded differently than regulatory scientists proceeded. All the scientists acknowledged the difficulties of extrapolating from their experimental work with cell lines or animals to the human population. The clinical scientists' work was usually based on clinical findings, and it aimed to explore unresolved questions that had been overlooked in mainstream science. Alternatively, it brought separate bodies of theory together that had not previously been linked. Scientists engaged in basic research emphasized the importance of collaboration for the progression of their work. Those working on the development of cancer therapies also stressed the importance of collaboration. However, collaboration with the private sector seemed to be more significant than in the past, because the drugs being administered were supplied by the pharmaceutical industry; whereas government funding had previously subsidized the development of hormones used in research.

The scientists I spoke to who were involved in basic or medical research related to IGF-I had not been aware of the controversy surrounding rbGH while they were conducting their own investigations. They had not considered questions about the safety of rbGH because their own work was not motivated by the controversy. However, in conversation none of the scientists thought it likely that rbGH posed a human safety risk, either because they assumed it would be broken down during digestion or because even at elevated levels the amounts of IGF-I in milk would be too small to cause systemic effects.

Although they did not expect the use of rbGH to pose a problem, they thought that this needed to be demonstrated empirically and that different kinds of experimentation from those actually conducted were necessary to prove this. A university research scientist who had identified IGF-I as a growth hormone mediator in the 1970s did not anticipate that the use of growth hormone would lead to adverse human health effects. However, he added that "if there is a problem, this can best be determined by long-term studies of animals reared on milk from cows treated with rbGH. Such animals should be studied to see whether they have abnormal growth or a higher incidence of teratological effects [birth defects] or cancer. I don't expect to see any adverse effect or any other detectable difference from control animals; however, that's the only kind of study that would address such concerns, since it would gloss over mechanisms and tell us what is the net effect."

Another research scientist thought that further experimental work needed to be done in order to demonstrate the validity of the assumptions about protein digestion and absorption. The scientist said that for years it had been thought that large proteins were degraded in the acidic environment of the stomach before they even reached the rest of the gastrointestinal tract. However, that "turns out not to be the case; all these peptides are extracted in acid, they're very acid-safe." When, after passing through the stomach, they reach the small intestine, he would predict that they would be degraded by digestive enzymes, but he was "not sure that anyone has looked at it carefully and *shown* that they're degraded."

The scientist posited that the FDA's measurement of gross organ weights was not sensitive enough to measure a growth response in the gut lining; the agency's methods would only detect massive changes. "Measuring organ weight is not a sensitive enough method to look for a trophic [growth] effect. You've got to take the tissue apart and examine it. They would never have found the answer they were looking for by weighing gastrointestinal tissue; they'd have to get a massive response."

In the "Freedom of Information Summary," the FDA stated that on microscopic examination, the gastrointestinal tract of orally treated rats was not different from that of normal rats (FDA 1993d, section 7c). However, the details of the experiment were not included in the summary. In order to accept the FDA's conclusion, which differed from that of his own studies, the research scientist wanted to see more information about how the experiment was conducted and what methods of measurement were used; however, this information was not publicly available.

Not only are regulatory scientists unable to share information freely with the general scientific community, because of the proprietary nature of the data, they are also conscious of the impact of their findings on their corporate clients. One reviewer stressed that when data are submitted, there is no company input into the decision-making process: "They may come and bring you additional information, but they are never involved in any decision about the adequacy of the study, or the interpretation of the study." However, regulatory scientists operate with an awareness of the impact of their decisions on the drug manufacturer, and they are reluctant to demand further testing unless it is absolutely necessary, because of its potential impact on the drug manufacturer. The FDA found it necessary to develop new guidelines for trials with protein hormones, not only to improve the accuracy of the scientific information but to avoid requesting more studies from the companies when initial studies proved irrelevant to the reviewers' concerns.[7] "When we did go back to the companies and tell them we had to do another study they were not very happy. In fact I thought they were going to slit my throat." One scientist distinguished between "reasonable" and "unreasonable" requests for further studies. Although it was believed that reviewers had to be assured about the validity of their decision, the scientist also understood companies' frustration with the decision-making process and believed that requests must be kept within reasonable limits: "I think it's unreasonable for the companies to say you can't keep coming back and asking us for information. But I understand their point of view because in some cases I think they feel it's hopeless; they're always going to be asked for more and more. I'd say if a scientist is being reasonable they are justified in requesting any additional studies. If, on the other hand, they're just off on their own thing, it's up to the supervisors to deal with that situation."

The scientist also perceived that the IGF study would not be a useful one for the companies to carry out, because at the time, IGF was not readily available, and so the companies were therefore going to have to produce it themselves for use in the clinical trials. Because there were four companies vying to get to the marketplace first, producing the IGF-I and conducting the clinical trials was perceived as an additional competitive pressure.

Monsanto, however, claimed that it was prepared to meet the regulatory guidelines. According to a Monsanto spokesperson, the company was able to direct its resources to meeting regulatory requirements because its personnel were not diverted by other biotechnology products. A Monsanto scientist also said that based on the company's review of the literature, it had anticipated the need for IGF-I studies before the FDA's request.

Other Human Health Concerns

IGF-I has been the major human health concern. However, the nutritional composition of milk and other consequences of the introduction of the drug were also an issue. Juskevich and Guyer (1990) have concluded that rbGH treatment did not have a significant impact on the nutritional quality of milk. Kronfeld (1989, 288) and Mepham (1992) have questioned the conclusion that milk from treated cows has the same nutritional value as milk in its current form. Mepham (1992) and the Toronto Food Policy Council (City of Toronto Board of Health 1991, 4) have also argued that rbGH could have negative consequences for public health if the public reduced its dairy consumption in order to avoid products from treated animals. Mepham's concerns about dairy consumption were based on European Community surveys of consumers' attitude toward biotechnology, in which respondents reported that they believed genetic engineering may involve risks to health and the environment and that they did not trust information about biotechnology that came from industry. Since most of the information about rbGH did come from industry or industry-sponsored sources, Mepham expected a consumer backlash against milk if the product were licensed. The rejection of milk would lead to reduced calcium and nutrient intake – possibly effecting an increase in osteoporosis – and substitution with less nutritious beverages containing more sugar (1992: 738).

Dr Michael Hansen from the Policy Institute of the Consumers' Union also raised the possibility that the approval of the drug could hasten the spread of bovine spongiform encephalopathy (BSE), also known as "mad cow" disease. Hansen claimed that treated animals require more energy-dense food than those not treated and that rendered cattle were used to supplement the energy-and protein-density of feed. Hansen acknowledged that there were no reported cases of BSE in the United States but argued that the existing BSE surveillance plan may not be identifying the population most at risk.[8] In the United States a hundred thousand cows simply keel over and die every year for no apparent reason ("downer cow" disease); Hansen pointed to a study from which it could be inferred that this disease may be a variant of BSE.

ANIMAL HEALTH

The major animal health issues were mastitis, a bacterial infection that causes inflammation of the cow's udder; reproductive problems; and lameness. By the time the drug had been approved in the United States, it was almost universally agreed that rbGH treatment correlated with an increase in the incidence of these diseases.[9] What was controversial was the extent of the increase, whether it could be directly attributed to the drug, and whether farmers could manage it. Critics argued that the increase was an animal welfare, and hence an ethical, issue and that the management debate did not adequately address welfare concerns. The animal welfare argument was ultimately accepted in both Canada and Europe. Canadian scientists, however, also expressed reservations about the data, which made them extremely wary of safety claims by the manufacturer.

As noted above, animal health also has a human health dimension. Although this dimension was publicly examined in the United States, the scope of the question was limited to antibiotic residues in milk. In Canada, however, dissenting scientists have been more concerned about increased antibiotic resistance, partly because this issue has become a greater public health concern since 1992. The human health implications of animal health issues will be discussed in the next section.

The Regulatory Review

As in the case of the human health review, existing scientific knowledge was important in assessing the animal health data.[10] Like the reviewers of the human health data, reviewers of efficacy and animal health data relied on their training and scientific knowledge when deciding what kinds of studies should be undertaken.[11] Reviewers typically have PHDs in animal science and/or statistics or doctorates in veterinary medicine. Graduate school training includes research in their particular areas of expertise, such as animal nutrition, reproduction, physiology, or genetics. Veterinarians at the Center for Veterinary Medicine (CVM) gain clinical experience during their training. Centre scientists and veterinarians also continue to attend scientific and/or veterinary conferences in their areas of expertise to remain current in their knowledge.

However, reviewers' knowledge of dairy farming from sources other than conventional science was also important. Reviewers drew on their experience with investigational herds during their animal science training or previous employment. Some reviewers had also gained scientific/veterinary knowledge through employment before joining CVM. Others had additional farming knowledge from their own experience of growing up on a farm and from their experience with investigational herds during their animal science training or previous employment. One scientist also reported that they relied on their knowledge of dairy farming to form judgments about "what farmers would need to know about the effect of such a product on their animals."

Although the FDA had experience evaluating production drugs for other species, it initially did not have extensive experience assessing dairy production drugs. Consequently, FDA reviewers obtained advice from the companies and from academic scientists regarding the evaluation of rbGH. One official explained that "Deciding how to record and analyze health data, e.g., daily health observations, was challenging because dairy scientists tended to perform less research in this area than, for example, [in] milk production, nutrition, and reproduction." The evaluation of Posilac was regarded as "a learning process" for the FDA, in which the reviewers' ideas might be modified as results came in and were evaluated. This gained knowledge then had an impact on the design of subsequent protocols submitted for future investigational studies for dairy production drugs. Also, since the dairy studies were conducted over longer periods of time, there was an unprecedented amount of data to analyze.

The learning process took place through interaction with the companies, who were able to challenge the reviewers' requirements. The FDA accepted a company's recommendations if they were scientifically justified and still allowed a valid evaluation of the investigational drug. An FDA official argued that "They're the ones doing the research out there, they might be a little more current on things, and if they can come in and present a solid argument for doing something differently than we recommend, then we'd think about it, and if it makes sense we'd accept it. So we learn from the companies as well." Monsanto's interaction with the FDA began when the protocols for the trial were developed and continued until after the drug had been approved and a postapproval monitoring program (PAMP) had been established. The company set up a meeting with the FDA to debate the issues ninety days after the protocol was submitted. Company scientists could argue against the FDA's recommendations or suggest that an aspect of the studies be conducted differently. According to a company scientist, "99 percent of the time we accept what they're suggesting but sometimes they haven't thought about some of the issues and so they may modify it after they understand why we were asking for that specific parameter." One example of protocol change was in reproduction; at the company's suggestion, cows continued to be bred beyond the usual measurement period, which showed that cows could still reproduce;

had the initial measurement period been maintained, the results would have been less favourable. Another example was in mastitis measurement – the protocols did not include microbiological sampling in the pretreatment period until it became apparent that Posilac-treated cows had a higher mastitis incidence, so the company wished to determine whether this was caused by the treatment or whether the treated animals were already predisposed to the disease. This allowed the company to argue that treated cows had been predisposed to the disease and therefore that mastitis could not be attributed to the drug. A scientist expressed this relationship as "an exchange of scientific opinions, typical of what occurs in any type of scientific organization. However, the FDA does have the final decision, and so while a company is certainly permitted to argue at length about any of our requirements, if we're not satisfied with their justifications, we don't accept their argument."

When the trials were completed, company scientists had "many" conversations with regulators, especially on the mastitis issue. After the initial data review, the FDA had concluded that there was an increase in mastitis incidence; the company then had to determine the extent of the increase, the impact on the dairy herd, and whether the increase was "manageable." The companies were informed of what specific level of increase in mastitis incidence would be of concern to the FDA. (The FDA would not reveal this information, because it was a preliminary decision based on dairy production drugs evaluated to date and currently accepted concepts on mastitis incidence in the dairy industry, which could change over time). Company scientists were concerned that the warning label that accompanied the product would overemphasize mastitis, and this issue was discussed with the FDA. The company was happy with the outcome of the discussions and felt that the mastitis issue was "in perspective."

The FDA communicated frequently with the company during the data evaluation and during consideration of labelling language. Ultimately, the decision on product approval and labelling was made by the FDA; however, the company was consulted after the data had been evaluated and before the labelling language had been determined. An official noted that:

> The FDA allows firms to discuss agency conclusions. When we review something and draw a conclusion, we relay that to the firm and let them respond. There are certain health factors that we've concluded are significantly affected by the product and should probably be on the label; the company might have an idea that the observation may be related to something else. They may have a better way of expressing it [on a label] to better communicate that to the user. Or they might have a valid reason for suggesting it not be on the label, and if

they have a sound, supportable reason not to put it on a label we would consider it and if they can convince us we will accept it. So there is certainly communication with the firm; I don't want to imply that we don't communicate with them a lot.

The centre also consulted with animal scientists in academia. In an interview it was explained that "We're free to contact anyone as long as we're not relaying proprietary information." Although the reviewers could not speak specifically about a particular product, it was commonly known that rbGH was under review: "Everyone in the dairy science community knew ... because the companies had been very public." In an FDA scientist's view, consultations with the academic community helped to fill gaps in the reviewers' knowledge and to keep them updated on scientific developments: "While I have a pretty good understanding of all areas of dairy science, my training had been in nutrition and metabolism, and I was not as strong in reproduction and health, so I might call professors that I know in dairy science with expertise in this area and say we are evaluating production drugs in dairy cows and what are the important things we should look for, we're trying to decide what needs to be looked at in these studies."

However, Health Canada demonstrated a more skeptical attitude toward the interpretation of scientists outside the Bureau of Veterinary Drugs (BVD). Although a scientist formerly with Health Canada's BVD had conversations with organizations and individuals outside the bureau, he argued that the review was "strictly a Health Canada decision." He had spoken to FDA representatives and to principal investigators at university sites within the United States, but claimed that these conversations did not influence his view of the data submitted. During the review he kept abreast of the literature, particularly in the *Journal of Dairy Science*, which he regarded as a reputable source of information. However, he differentiated between the results reported in the literature and those he had seen in the Monsanto submission: "If you read something, you don't automatically believe it. You file it in the back of your head and you think, well, [this scientist] had these results, someone else had other results, and we're looking at a set of data that maybe contains both sets of results, but they [the company] didn't follow the protocol ... You come to a conclusion based on what the company has done, not what somebody else has done ... They have to be able to demonstrate to me based on what they did that it's safe and effective, not what Joe Blow did in South Carolina."

The studies evaluated by the FDA's Center for Veterinary Medicine were conducted by Monsanto at two sites in Missouri and at several land-grant universities in the United States: Arizona, Cornell, Florida, and Utah.[12] An efficacy study was also conducted in Idaho and an injection-site reaction study in Vermont. (Studies must be conducted at at least three different geographical locations). Once the Investigational New Drug Application had been approved, the company submitted a New Animal Drug Application (NADA), and further testing took place to determine drug safety and efficacy. Studies in support of this application evaluated the company's claim that the use of the drug would increase the production of "marketable" or "saleable" milk, that is, milk that does not contain drug residues (such as antibiotic residues) above the FDA-established tolerance level (FDA 1988).

In judging what kind of trials should be requested, the FDA reviewers sought a balance between obtaining scientific information and allowing for experimental ease. "You have to make a happy balance where you're getting the information you need, that will be of practical importance to the users of the drug, but you're also not making the study so complicated that it's impossible to run." Since several variables were being examined in each study, reviewers had to ensure that the collection of data on mastitis, for example, would not interfere with gathering or interpreting data on reproduction or vice versa.

Another consideration was ensuring that the trials reflected actual dairy practice. However, actual dairy practice in the United States included the use of drugs for the regulation of reproduction and treatment of infection, which is common in large-scale, industrial farming. Therefore, the FDA accepted data from clinical trials in which animals were treated with "extra-label" drugs, that is, drugs used in ways that violated the FDA's guidelines. For example, a medication designed for another species may have been used on a cow, or a bovine drug may have been given in larger-than-recommended doses.[13] Thus, the FDA accepted studies for Monsanto's product, Posilac, in which the use of extra-label drugs was considered consistent with accepted practices on U.S. commercial dairy farms. However, Health Canada did not agree with this decision. According to the gaps analysis report, the company had failed to show that milk production would increase when a farmer used only approved drugs (Internal Review Team 1998, 13). In the United States, the use of the sex hormones prostaglandins and gonadotropins, which are commonly used to restart reproduction after a cow has calved, was permitted in clinical trials, usually only after a specific waiting period. During the waiting period, a more direct influence of rBST on reproduction could be evaluated. According to an FDA scientist, "we wanted a controlled study, but we also wanted to be realistic, to reflect the real world, so this was a way to balance that."

The FDA perceived its "ultimate customer" as the consumer and producer of food. However, the reviewers' second concern was providing the firm with a fair and timely review. Reviewers were conscious of the importance of responding to company submissions in a timely manner, which also affected their career advancement: "Part of our yearly performance evaluation as reviewers is our timeliness." The centre maintains a tracking and reporting system that keeps track of all submissions by companies and their "due date"; the amount of time assigned for review is dependent on the nature of the submission. Reviewers can obtain a listing of their pending reviews in order of the due dates. "We try to work on things that fall due first. We're definitely very conscious of those things." Protocols got a short review time – forty-five days. "They're given priority because you don't want to hold up a company's opportunity to start a study." No contradiction was perceived between the FDA's regulatory requirements and company interests. Requesting further studies or not approving a product may benefit the company as well as customers because it prevents the introduction of unsafe or ineffective drugs: "If something causes serious problems down the line, it will hurt the company's reputation."

Animal safety data were compiled from several sources. Traditionally, the FDA required that companies conduct two key studies to determine the safety of an animal drug: an acute toxicity study, a controlled study in which a small number of animals receives up to twenty-five times – this was later reduced to ten times – the expected dose for up to fourteen or twenty-eight days; and a chronic toxicity study in which one, three, and five times the highest expected dose is administered. The latter is also known as the 1, 3, 5X study. By giving exaggerated doses of the drug, the fourteen-day study is intended to highlight any potential safety problems, which can then be investigated further in other safety or efficacy studies.

By 1988 the data from the first set of Monsanto safety and efficacy trials had been submitted.[14] Routine toxicology testing did not identify any specific effects for rbGH; instead, common ailments became more prevalent (Kronfeld 1994, 116). In response, the agency decided that further safety data should come from observations and milk sampling carried out during the efficacy studies (FDA 1988, 1). The protocols for both the efficacy and animal safety studies specified that each animal should be examined every day and any unusual health observations recorded. According to an official, the most valuable animal safety data came from the efficacy studies, because animals in efficacy studies are managed under conditions more likely to be encountered at commercial dairy farms. Much more animal health data was obtained at doses of the drug likely to be approved, because of the larger number of animals used in efficacy studies. As a consequence, the 1, 3, 5X study is no longer required for production drugs.

The animal health critic Dr David Kronfeld has stated that the evaluation of health observations from efficacy studies could not adequately determine the safety of the product, because efficacy trials were not specifically designed to examine safety variables (1994, 116). The Canadian animal scientists Burton et al. have also argued that the failure to investigate safety as a primary objective of the long-term studies has meant that "Adequate statistical analysis and interpretation of the resulting reproductive and health data have been difficult or impossible due to the small number of cows on individual trials" (1994, 178). Former FDA veterinarian Dr Richard Burroughs, who was dismissed and then reinstated by the agency, also argued that the undertaking of health observations in efficacy studies was not adequate to assess the impact of the drug (Burroughs 1994, 13–14).

Because there were not enough animals in individual trials to draw reliable conclusions about the effect of drug treatment on the incidence of infection, particularly mastitis, the FDA analyzed the data from individual trials and pooled the data from eight trials.[16] Critics, however, contested the pooling of the data. The most outspoken animal health critic, Dr David Kronfeld, has acknowledged that an increased mastitis frequency has not been observed in all herds but in "1 of every 2 or 3 BST-treated herds" (1994, 123). He has argued that pooling of the data obscures the severity of mastitis incidence in herds where it is a problem.[17] Kronfeld relied on studies conducted at Cornell and Vermont, a collaborative report by White et al. from the principal investigators on the Monsanto trials, and a report by Thomas et al. reviewing data from fifteen U.S. herds.

The companies submitted their analysis of the results before submitting the data package to the FDA. Once the analysis had been submitted, the FDA requested that statistical analyses be done according to particular designations. If there was an increase in a particular disorder, the regulators requested further analyses to determine what was affecting the disorder. When the analysis had been completed according to these specifications, it was forwarded to the FDA with the electronic data and files, at which point FDA statisticians re-analyzed the data.[15] If they were not able to duplicate the company's results, company statisticians were contacted to determine whether the numbers had been entered incorrectly.

When animal health observations were reported at the trial site, the FDA required that the observer record the date, time, his or her name or initials, the animal ID number, and the nature of the observation. If an abnormality was reported, subsequent treatment was also recorded. This initial documentation is known as "source" or "raw" data, which is transcribed into a data-base at the firm. The company and the principal investigator are expected to maintain this information safely. When auditing the study results, the FDA may request that the sponsor submit a copy of certain portions of the "raw" data, or they may examine the originals when doing an inspection at a study site and/or of the firm's data base.

Both FDA and Health Canada reviewers reported concerns with the accuracy of the recording, summarization, and analysis of animal health data. An FDA official noted that the CVM now asks for much more detail in the sections of the protocols describing proper data collection procedures and in the examination of pivotal data, because of problems with the authenticity and/or accuracy of data submitted by the companies: "The fact that we are very careful in that section of the protocol ... there were reasons for us to be that specific. Maybe it was this company [Monsanto] maybe it was another company. I can't be specific, but it's the fact that we are very careful with that section of the protocol." It would appear, therefore, that the centre was able to detect when data were not completely accurate.

In Canada scientists also had concerns about the accuracy of company data. According to a former Health Canada scientist, the data submitted by Monsanto did not conform to the company's own protocols, and this caused him to question the acceptability of the data. Before receiving the data from Monsanto, the BVD's evaluators came up with a list of basic requirements they expected to see addressed by the data. They had also been reading articles from refereed journals regarding experimentation in other countries and may have added requirements based on that knowledge.

When the scientist received the data, his main concern was determining whether the data conformed with the original protocol that was included in the package. "One of the first things I did was to read the protocol, to find out if they had actually followed it. Did they do what they said they were going to do? Now, I won't go into the details, but the bottom line is that they did not. And what they did not do I won't discuss with you. Basic science asks, if you didn't follow the protocol, why not? Where was the explanation? This was not provided. There was a real concern with the company with regards to [animal] safety ... and I discussed this with the company, both in meetings and by letter." He was disturbed by the company's failure to conform to its own protocols: "There were things happening in those experiments that were just horrendous. Now I can't get into that because that's privy information, but in my opinion, it was very bad experimentation. I had a real concern that those experiments were not done as they should have been done. I corresponded with them about this and they said fine, we'll come in and talk to you, and get this straightened out. Well, it never did get straightened out." The scientist asked for raw data from the sites to compare with the company's summary reports and was not satisfied with his investigations; "I didn't have to go very far through all these sheets ... before something told me this is not good."

According to the scientist, Monsanto agreed with his criticisms of the data. He reported that he had discussed the Health Canada decision with Monsanto representatives at a meeting in the fall of 1995. Five Monsanto representatives, mainly from the U.S. parent company, were present at the meeting, along with the assistant deputy minister of Health Canada, the director general, and the director of the bureau's Human Safety Division. The company representatives agreed with the bureau's analysis of the data. "When we finally sat down with them and said, this is wrong, they agreed."

Company scientists, on the other hand, emphasized the veracity of the data and the rigour of the data-checking process both at the company and at the FDA. They presented a different picture of the review process in Canada. They perceived it either as unduly influenced by political concerns or as merely delayed because Health Canada had fewer resources to devote to the evaluation than the FDA. A Monsanto spokesperson described the Health Canada decision-making process as similar to that of the FDA:

Basically, it's the same insofar as they base all their decisions on science, like the FDA. The outside pressures can slow things down, but at the end of the day they don't change the science – the [FDA] decision was based on science and that's true in Canada too. I've been interacting with Canada since 1985. Just letting them know what the requirements were. They were a little further behind in what should be required for this kind of product. As a company, we really focused on the U.S. to get approval. We did do some work up in Canada, but it was for a product form we didn't go forward with. One of things that's different is they have fewer people to work on it, and I think they were overwhelmed with the amount of material and being able to manage that.

In fact, the representative indicated that the process was not taking any longer in Canada than in the United States, because the Canadian submission was made about three years after the American submission:

We're still at the same point in Canada similar to the time it took to get through the FDA. All our effort was to get U.S. approval. It's the key approval, to help you get approvals in other parts of the world. U.S., Canada, and Europe – if you get any of those approvals it helps you anywhere else in the world, in South America, South East Asia. When we had the issues in Europe we put all our focus on the U.S. to get approval first. We didn't put the resources in Canada. It's just taking a longer period of time. The FDA has a lot more depth. They reanalyzed all the data, they had computer know-how and power to do that, where Canada has not had that, and resources to physically look at the files.

Both Health Canada scientists and Monsanto regarded the request for a voluntary moratorium as irrelevant to the review process. According to Monsanto representatives, the company knew that approval was unlikely during the next twelve months, so the moratorium made no difference to their strategy. A Monsanto representative commented that "We weren't expecting approval at all. They wanted to set up a task force and didn't want approval before the task force was finished. We weren't expecting approval anyway. That was an artificial date of a year's time so they could complete the work – it wasn't unexpected."

A former Monsanto scientist who had been involved in the international marketing of bST had a different perception of the company's interactions with Health Canada and stated that Monsanto was extremely frustrated when approval was not announced when it was over: "At the end of that it became clear also that there wasn't going to be any serious intent by Health Canada to approve the product. So we basically withdrew and weren't going to spend resources on it any more ... We no longer have an office for this product in Canada. The registration's still active, but I can't comment on that, I haven't been with the company for nine months. We pulled our people out of the office there and decided we weren't going to spend additional resources the way we did before because it was not viable to do so. We were given no assurances at all that people [at Health Canada] were working in good faith." The scientist emphasized that since he was no longer with the company, he was not aware of the current situation. However, he felt that, as in Europe, sociopolitical influences rather than science-based arguments influenced the process in Canada.

According to a Monsanto official, on the other hand, Monsanto had decreased its commitment to obtaining Canadian approval in order to concentrate its resources on the U.S. process. It was also indicated that the company had encouraged interaction between the two regulatory agencies: "We have tried to interface the two agencies because there's NAFTA. It doesn't make sense – both the CVM [the FDA's Center for Veterinary Medicine] and the BVD [Health Canada's Bureau of Veterinary Drugs] are being cut back; why are they duplicating this work? We're trying to get them to interface and use each other's knowledge. There were two people from the BVD at both VMAC meetings, and we've given authorization so there are no confidentiality problems between regulatory groups. They have access to either set of files."

A former Health Canada scientist also acknowledged that he had

had conversations with officials in the United States, but would not divulge the content of those discussions.

In reaching conclusions about the results from both individual trials and the pooled data, reviewers distinguished between the "statistical" and "biological" significance of variables affected by treatment of animals with the investigational drug. As with judging what kind of studies would be most appropriate, scientists drew on their background and training to make this distinction. The current literature and the results of their inspections were also important in reaching a decision. Judgments were made collectively; each reviewer's concerns were discussed by the group.[18] The information that appeared on the product label reflects what the FDA felt were not only statistically but also *biologically* significant results, in which the effects "were seen consistently and in conjunction with other variables pointing in the same direction" and could, in their judgment, be attributed to use of the drug. The FDA decided that three areas of animal health were clearly affected by rbST: reproduction, mastitis, and nutrition.

Once conclusions had been reached about the biological significance of the incidences of infection, reproductive problems, and foot disorders, reviewers then decided whether the problem was severe enough to render the drug unapprovable or if it could be dealt with by informing the producer through product labelling. Problems that were regarded as "severe" included death, painful conditions, or conditions that were not necessarily painful but that were increased dramatically compared to controls – "if in control animals it's something you see sporadically, but in treated animals just about every cow has this problem."

Severity was also determined by the novelty of the ailment and was linked to the concept of "management." Since mastitis, for example, was a common disease among dairy cattle, evaluators considered whether this problem could be "managed" by the farmer administering the drug if the FDA decided in favour of approval. "What we saw didn't suggest that [mastitis cases associated with Posilac use] were harder to treat or that they were [caused by] different types of organisms ... so it wouldn't all of a sudden cause a new problem that no-one knows how to handle." Since the difference in mastitis incidence associated with rbGH had been determined as less than the difference between, for example, early and late lactation, it was reckoned that any additional risk could be managed. The management concept also related to farmers' skills in handling health problems; since it was accepted that farmers were already managing mastitis or reproductive problems associated with high production or environmental factors, it was inferred that any additional risk associated with the use of the drug could also be managed. According to an FDA scientist, "your early lactation cow is the best example. That's when she's most prone to mastitis. Successful dairy farmers are obviously handling that, so in later lactation when they're at less risk for mastitis, you add BST on top of that, they're going to be able to handle it, in our judgment." The reviewers decided that it was important "to let farmers know that the use of Posilac could exacerbate those problems, if you already had problems in those areas. We decided that the best way to express it was to make sure you've got a good management program in place."

Health Canada disagreed with this conclusion, as did the Canadian Veterinary Medical Association (CVMA), which felt that although current dairy health management techniques "could reduce this increased risk, they are not adequate to eliminate it" (CVMA 1999, 2) and that therefore there were legitimate animal welfare concerns raised by the use of rbGH. The CVMA concluded that use of the drug increased the risk of clinical mastitis by 25 percent and the risk of clinical lameness by approximately 50 percent, and that it reduced reproductive performance as well as the animal's lifespan.

Health Canada regarded the labelling issue differently than its American counterpart. Canadian animal drugs have a warning label to protect human safety and a cautionary label to indicate animal safety problems. The cautionary statement indicates any animal health problems that do *not* justify the refusal of a Canadian license. According to the scientist, the reproductive problems, mastitis, and other "major issues" were too significant to be dealt with on a cautionary label: "Our system says you have to demonstrate that it's safe and effective, and that means that you can't come out with a label that says you can use this, but beware of all these things that could happen. That's not right, in my view." He did not regard the U.S. approach of labelling to indicate potential animal health problems as acceptable in Canada but could not give a detailed response to the FDA's freedom of information summary.

In Canada, the definition of the concept of "safety" was more stringent than that of the FDA reviewers. If around half or more than half of the treated animals showed signs of illness, this was regarded as highly problematic. If fewer than half the animals were ill, he would want to obtain further evidence that this level would not increase: "Suppose with BST in less than 1 percent of the cases you see a certain type of mastitis. Well, that's not enough to say we can't market this drug. Statistically that's very low, and you make that statement on the label ... If you've got 50 percent or more, that's an automatic no. With 50 percent or less, if say 48 percent are getting mastitis, you say to the company you want more studies done to show that it remains at 48 percent, or perhaps that it's going to drop. If it stays up there it's no."[19]

Mastitis was seen as a serious issue, however; even a 10 percent incidence in treated animals would cause him some concern. The mastitis issue was viewed in the light of Canada's supply management system. If a farmer has to throw out milk because of high somatic cell counts or antibiotic residues, he or she would have to buy milk to make up the quota or pay a fine.

Health Canada's position was similar to that of the critics in the United States, who argued that the disease was not "manageable" in treated animals. For Kronfeld this was an animal welfare issue; regardless of whether mastitis is an indirect human health problem, it is a painful disease for animals that is already difficult to manage and prevent, and the introduction of a drug that increases its risk is unjustifiable. In Kronfeld's view, the FDA-Monsanto judgment that BST-induced mastitis is manageable by current methods is not supported by the data.

Monsanto scientists, on the other hand, took milk yield into consideration as a factor that may affect the variability of mastitis. They distinguished between the "direct" and "indirect" effects of the drug. For example, company scientists were aware that there is a relationship between high milk production and mastitis; high-yielding cows tend to have a higher incidence of disease. A higher incidence of the disease in rbGH-treated cows, could, therefore, be attributed to increased production rather than to the direct effects of the drug. In an article coauthored with the principal investigators of the product trials in the United States and Europe, they argued that the incidence of mastitis in treated animals was similar to that of the incidence in cows with naturally high levels of milk production. Therefore, when mastitis incidence was examined in relation to milk production, "sometribove [rbGH] had *no effect* on the incidence of clinical mastitis" [my emphasis] (White et al. 1994, 2250). In coming to this conclusion, the researchers reported evidence of higher mastitis rates in cows genetically selected for milk yield, and observed that this relationship was not altered in treated animals (2256). It was also noted, however, that the association between milk yield and mastitis is small and does not have a major influence on the incidence of mastitis (2250).[20]

University scientists tended to share the company scientists' perspective on animal health issues. The principal investigators for the trials distinguished between "catastrophic" and "subtle" effects on animal health.[21] A university scientist explained that mastitis incidence was evaluated from a number of perspectives but that examining it in relation to milk production was most meaningful for evaluating food safety: "In terms of food safety issues, the concern about mastitis isn't per cow, it's per unit of food produced. So the reason why we looked at it that way has to do with the food safety issue. If I've got a cow that only has one case of mastitis but gives twenty pounds of milk and another one that gives ten times that and has two cases of mastitis, the chances that the drug could lead to contamination of the milk by antibiotic residues are very low ... any risk factors are related to cases per unit of milk." He added that "If herd A averages one case of mastitis and 12,000 kg of milk per cow per year while herd B averages the same incidence of mastitis per cow but only half as much milk per cow, the risk from inappropriate drug use (e.g., antibiotic residue) is less per unit of milk in herd A." Other animal health parameters were also viewed in terms of their relation to milk production.

This method of analysis was not acceptable to the FDA, however. Regulatory scientists regarded the cause of the effect as irrelevant. According to a Monsanto scientist, the FDA did not accept the indirect-effect argument but "looked at how much milk would be lost, how much would be spent on antibiotics, and they determined that it wasn't going to significantly affect profitability on the dairy. So they determined it was approvable. That was before the external review."

ANIMAL HEALTH AS A PUBLIC HEALTH ISSUE

The possibility that increased rates of mastitis would result in increased antibiotic residue in milk and antibiotic resistance in the general population was first raised by Epstein in 1989. It became controversial as a public health issue when the General Accounting Office released a report arguing that the FDA had not considered this potential indirect risk to public health from antibiotic residues in milk (GAO 1992b).

In response to the GAO, the FDA had tentatively concluded that the indirect public health risk was insignificant (Guest 1993, 2); however, it convened a one-day public meeting of its Veterinary Medicine Advisory Committee (VMAC) to address the issue. The FDA asked the committee to consider whether the use of the drug resulted in a "meaningful biological effect" on mastitis; whether when compared to other factors that influence the incidence of the disease, the increase exceeded an acceptable threshold and could be managed; and whether its use would contribute significantly to illegal drug residues entering the food supply. The questions were framed in such a way that the answer was determined by the question. The second question asked the committee to determine whether the mastitis increase exceeded "an acceptable threshold," but the phrase "when compared to other factors" prescribed how that threshold would be determined in relation to other factors identified by the FDA. It was the validity of this comparison that had been questioned by the critics. In order to reach a different conclusion from the conclusion reached by the FDA, one had to dispute the comparative method of analysis. For the centre the relevant questions were whether the increase was "meaningful" and whether it was "manageable" in two senses: could the animal health risk be managed by farmers without becoming an unacceptable practical or economic burden and could the risk to public health be managed by the existing system for monitoring milk drug residues?

In the FDA's view, since the risk of antibiotic residues entering the milk supply was being successfully managed by a national and state system of milk monitoring, the risk associated with rbST could also be managed by that system.

At this meeting, FDA officials emphasized recent steps taken to improve drug residue monitoring. The most commonly used drugs in the treatment of mastitis were beta-lactam antibiotics, which may cause allergic reactions in people who are already sensitized to penicillin. However, the FDA argued, the dairy industry had been required to test every milk tank for beta-lactams since 1992. Even if the residues were not detected, ingestion of residues in allergic individuals tended to result in a skin rash and was not a serious enough reaction to pose a significant health risk (Mitchell 1993, 7). The Consumers' Union, represented at the hearings by research associate Dr Michael Hansen, expressed concern about the potential for the spread of antibiotic resistance and antibiotic-resistant food-borne infections, particularly given that antibiotics were used illegally in Monsanto trials. The FDA did not discuss the potential for increased bacterial resistance to antibiotic drugs as a result of rbGH introduction. However, Monsanto's vice-president, Dr Virginia Weldon, pointed out that only 10 percent of antibiotics given to animals are used for therapeutic purposes; the other 90 percent are administered at subtherapeutic levels in animal feed to encourage growth. There was no conclusive proof that subtherapeutic doses of antibiotics presented any risk to human health; therefore, the use of rbGH would also not represent a risk (Weldon 1993).

There is still no conclusive proof about the impact of the routine administration of antibiotics in animal feed on human health, although there is some evidence that use of the animal drug avoparcin is partly responsible for the emergence of vanomycin-resistant bacteria in humans (Hawkey 1998, 1299). Since the approval of rbGH in 1993, there has been greater concern about the impact on increasing antibiotic resistance of subtherapeutic doses of antibiotics administered in animal feed as growth promoters. The EU has decided to act preemptively on this issue. In 1997 the European Commission banned avoparcin due to fear that it could hasten the spread of bacterial resistance (Soil Association 1998); and in 1998 the agriculture ministers of the European Union banned four feed-additive antibiotics, effective 1 July 1999. Although two reports written for the EU stated that "there is not sufficient evidence to identify the cause of resistance emergence" (Follet 2000, 154), the ministers decided to ban these products because of their relationship to antibiotics used in human medicine (Matthews 1998, 1). However, whereas the EU has banned the use of animal drugs that it fears may lead to increased bacterial resistance to human drugs, the United States has not taken similar action.

In fact, in the United States animal drug laws have been reformed in order to simplify and speed up the review process. The Animal Drug Availability Act of 1996 mandated that since the sponsor and the FDA have reached an agreement on testing requirements, it is binding upon both parties (21 U.S.C. §360(b)(3)). Another proposal that was relaxed was the "optimal dose" rule. Before 1996, the dose recommended by the drug sponsor could not exceed the level shown to be effective; for example, if both a 500 mg and a 750 mg dose were equally effective, the lower dose had to be chosen. Now, a higher dose is permitted, provided that the resulting residue level is still safe (see Lambert 1997). Clinton and Gore's National Performance Review proposed that, among other things, the evaluation system should be decentralized, so that the sponsor could communicate directly with FDA reviewers and so that the review of one part of the submission would not interfere with another part (Clinton and Gore 1996, 4). With regard to drug practices, the use of animal drugs at the veterinarian's discretion, rather than in accordance with the drug label, has also been codified.

In response to evidence presented at the VMAC meeting, the committee concluded that the product was approvable; that is, that "there is no risk, or the risk is insignificant." One of the VMAC committee members expressed the view that "we would have to wait 20 to 30 years to have good drug resistance data" (FDA 1993d, 5).

In Canada, as in the United States, regulators decided that antibiotic residue would be managed within the existing dairy system, which imposes harsh penalties for antibiotic contamination. Canadian milk is collected from a number of farms and pooled in a tanker, with samples of milk also taken from each farm. If a tanker is contaminated with antibiotics, all the milk is thrown out, and the farmer responsible for the contamination must pay for the entire amount, which may cost up to $18,000 (see the statement by Baron Blois, Senate Committee on Agriculture and Forestry 1998b). The antibiotic resistance issue was not debated until the Internal Review Team's "gaps analysis" report. The Expert Panel on Human Safety thought that this issue did not warrant concern given that the contribution of the drug to overall antibiotic usage in agriculture would be small.

At the Senate committee hearings, Dr Shiv Chopra argued that animal health was not an animal safety issue, but a human safety issue; he criticized the Expert Panel on Human Safety for not paying sufficient attention to it in their report. For Chopra, signs of illness in the animal should have led the human safety evaluators to ask further questions about human health. He made the analogy to forms of food adulteration that were prohibited in the original Food and Drugs Act and are still illegal. The act prohibits the production of food in a filthy environment, but rbGH treatment creates "a filthy milk production factory called the cow" (1998b). Neither the FDA nor Health Canada provided a rationale for their reasoning about the relationship between animal and human health until pressed to do so by external parties: the GAO in the U.S. case and the Senate in the Canadian case.

RBGH, SCIENCE, AND INTERNATIONAL RELATIONS

The European Experience

In Europe there were two separate assessments of the rbGH data, six years apart, that reached different conclusions regarding the acceptability of the drug. Although the first review of rbGH in the early 1990s favoured approval of the drug, by 1999 there was a regulatory structure in place in Europe that allowed for the expression of critics' concerns about it, along with a European Council directive that prevented the use of nontherapeutic drugs such as rbGH if they were "detrimental to the health or welfare of the animal" (Council of the European Union 1998, annex 18).[22] The scientific committee's conclusion that rbGH had an adverse impact on animal welfare, therefore, enabled the substance to be banned under this directive.

The initial scientific assessment of rbGH was made in the early 1990s by the European Commission's Committee on Veterinary Medicinal Products (CVMP), which, like the FDA, decided that the product was safe when used under specified conditions. Unlike the FDA, however, these conditions included veterinary supervision and a withdrawal period before slaughter and meat consumption (Levidow and Carr 1997, 31). Although this decision would have permitted the European Council of Ministers to approve the commercial release of the drug, the council in fact delayed its release due to concerns about its impact on the already massive European dairy surplus. The product was banned, pending further review, until 1994, when the ban was extended until the end of the millennium. By this time an alternative scientific assessment structure was in place in the wake of a crisis that had shattered the legitimacy of Europe's food safety system: mad cow disease.

In 1996 the British government had admitted the possibility of a link between a new variant of the degenerative brain disorder Creutzfeldt-Jacob disease and a similar illness in cattle, bovine spongiform encephalopathy (BSE), dubbed "mad cow" disease. The government and its scientific advisors had denied this possibility for approximately a decade before the announcement. But confidence in the government declined dramatically, and industrialized agricultural practices were scrutinized. Before 1988, when the British government banned the practice, rendered offal had been fed to cattle as a protein supplement. Although the material was treated at very high temperatures, the infective agent could survive the process. The government had taken steps to stop the feeding of rendered animal parts and to prevent the consumption of BSE-infected cattle; however, these measures were not adequately enforced, and random inspections later revealed the presence of infected meat at slaughterhouses (Powell and Leiss 1997).

The crisis undermined regulatory legitimacy not only in Britain but also in Europe. The public and national governments questioned the failure of the European system to prevent the crisis. At the European parliamentary inquiry into the issue, the agriculture commissioner raised doubts about whether existing scientific committees were sufficiently independent of member states and lobby groups (Butler 1995,503). Subsequently, the commission set up eight new committees to replace the existing ones and created a scientific steering committee to oversee their work (Rogers 1997b, 1823). The Consumer Policy Unit was granted responsibility for food safety (Rogers 1999a, 1529).

As a part of this restructuring, two new committees were assigned to evaluate the impact of veterinary drugs. The Scientific Committee on Veterinary Measures relating to Public Health assessed the public health impact of veterinary drugs; the Scientific Committee on Animal Health and Animal Welfare, as its title suggests, assessed their impact on animals. These committees undertook another review of rbGH in the late 1990s, before the Council of Ministers moratorium expired. The committees raised concerns similar to those that had been expressed in North America. Such doubts had been voiced previously in Europe; indeed, the work of British scientists such as Professor T.B. Mepham was important to the arguments of the North American critics, and there were ongoing exchanges of information between European and North American groups. Those critical of the drug, such as Mepham, were included on the animal health and welfare committee. In addition, the committees were reconsidering the rbGH issue in the wake of the gaps analysis report of Health Canada's Internal Review Team.

Like the authors of the gaps analysis report, the committee examining public health argued that several questions about IGF-1 had not been thoroughly investigated. For example, they were concerned that exposure levels had not been accurately determined, because differences in analytical procedure had not been taken into account. They asked whether other growth factors might be stimulated by rbGH and released into milk and whether fragments of the hormone could stimulate production of growth factor in humans. Although the IGF-1 level likely to be consumed was lower than in the human gut, given preexisting levels, what was the potential effect of an additional amount in the long term? The committee advocated an evaluation of the effect on the gut of infants, and on gut cancers (Scientific Committee on Veterinary Measures relating to Public Health 1999).

The committee on animal health and welfare reached conclusions similar to those of Health Canada, arguing that drug-induced levels of mastitis, reproductive problems, and leg problems were unacceptably high and that therefore the drug should not be used. Whereas the CVMP had acknowledged the risk of mastitis but accepted it as an effect of greater production, to undermine this earlier decision, the later committee emphasized the link between production and disease. The later review also criticized the introduction of an intermediate factor, production, to explain away the relationship between the causal factor, bGH, and its impact on the incidence of mastitis (Scientific Committee on Animal Health and Welfare 1999).

rbGH and International Trade: The EU/US Conflict at the Codex Alimentarius Commission

The conflict between the European and American positions on rbGH was played out at the international standard-setting body, the Codex Alimentarius Commission.[23] Codex, which was established in 1961, formulates food safety standards, including, for example, the acceptable levels of pesticides or animal drug residues in food. Standards are usually expressed as maximum residue limits (MRLs) (that is, the maximum level of a substance or residue that is allowable in food); or as an acceptable daily intake (ADI) (that is, as the amount of a substance that can be consumed daily without threat to human health). Because discriminatory trade practices may be justified by food safety standards that are more restrictive than those of a country's trading partners, Codex has aimed to facilitate global trade in food products through the establishment of international standards and the harmonization of domestic ones (Codex 1999).

Codex decisions became far more influential as a result of the Uruguay Round of trade negotiations, which recognized Codex standards as the norm for harmonizing national measures on food safety. The Agreement on the Application of Sanitary and Phytosanitary Measures (the SPS Agreement), which recognized Codex standards as providing the basis for harmonization, was negotiated during the Uruguay Round, and it entered into force with the establishment of the WTO on 1 January 1995 (WTO 1998). The SPS Agreement applies to all health protection measures, such as food safety standards, that may affect international trade (article 1 (1)). Therefore, although the agreement does not determine what standards a country may enforce on its own producers, it does affect a country's ability to ban imports on health and safety grounds if its standards are higher than those set by Codex.

Countries *may* implement standards more restrictive than those promulgated by Codex, but only if they can be justified scientifically (SPS Agreement, article 3 (3)). The legality of a trade-restrictive standard is determined by its scientific validity (at least in theory). If such standards are not deemed scientifically sound, the Codex standard will be taken as the norm in international trade disputes. In 1997 a World Trade Organization dispute-settlement panel rejected the EU's nine-year ban on the import of U.S. beef treated with (steroid) growth hormones, on the grounds that such a ban was not scientifically justified; Codex had previously found the three hormones in dispute to be safe (FAO 1997, 1). In the case of rbGH, the EU had not banned the import of milk or dairy products from the United States (COMISA 1997, 3), so the issue was not contested before the WTO. However, had a Codex standard been established, pressure would have increased on national governments to accept the safety of rbGH.

A proposed draft standard for rbGH was referred to one of Codex's scientific advisory committees, the Joint Expert Committee on Food Additives (JECFA). In its evaluation of the drug in 1993, JECFA found that when used in compliance with good veterinary practice, the margin of safety was so wide that there was no need to specify MRLS or ADIS (see JECFA 1999). Because Codex establishes food safety standards, JECFA discussed the animal health issue only in relation to its human health implications. The committee reached the same conclusion when it reevaluated the drug in 1998. By this time, the EU had raised several additional concerns about the human health implications of the drug.[24] The EU argued that although JECFA had found the drug to be safe, the committee's scientific analysis was too narrow to take into consideration issues of importance to EU consumers. Like the critics of the drug in North America, the EU argued that since rbGH had no therapeutic benefits, its risks should be analyzed more stringently.

It also argued that there were nonscientific issues that had not been considered in the rbGH case but that should be taken into consideration in the establishment of a Codex standard. In making this argument, the EU referred to Codex principles, which refer to both scientific and nonscientific factors: although the first principle concerning the role of science in the decision-making process states that standards and guidelines "shall be based on the principle of sound scientific analysis and evidence," the second principle allows for the consideration of "other legitimate factors relevant for the health protection of consumers and the promotion of fair practices in food trade" (Codex n.d.). Because these "other legitimate factors" have never been elaborated by Codex, the EU argued that a decision on rbGH should be postponed until a conclusion had been reached regarding them. According to the EU, "other legitimate factors" that should have been considered by Codex in relation to rbGH included the need to maintain consumers' confidence in the regulatory system, the potential impact of the drug on milk consumption, and animal welfare concerns (Codex 1997a, 1–3).

The United States, on the other hand, had consistently argued that risk and analysis and management should be "science-based" (Codex 2001, 2–5), a position that reflects the position of the animal drug industry, which has argued that Codex's failure to adopt a standard recommended by JECFA undermines the credibility of the agency and thereby threatens world trade and public health (Coalition of the Americas 1997, 30). The United States consistently voted for the adoption of a Codex standard for rbGH, and Canada has voted with the United States on this issue (see Codex 1997a, 1999a).[25] Although Codex standards are purportedly scientific, Codex procedures mean that proposed standards may be rejected by a majority vote of Codex members in spite of the recommendations of its scientific committees. In 1999, the EU successfully voted to postpone the discussion of rbGH, and the United States decided not to pursue the issue further at the Codex meeting in 1999, given that there was no consensus on the issue (Hansen 1999). Since that meeting, the issue has again been considered by a Codex committee. The EU has proposed that the issue should be taken off the agenda, since the Committee on Residues of Veterinary Drugs in Foods "should consider whether it wants to spend more time on the evaluation of a substance which has no therapeutic or preventive properties and whose use has been proven harmful for animal health and welfare and whose potentially harmful effects on human beings have not yet been fully elucidated" (European Community 2000, 2).

The rbGH case highlights the difficulty of adjudicating disputes over food safety standards by an appeal to "science." JECFA's conclusions were taken by the United States and Canada to represent sound scientific analysis, but these conclusions did not adequately address two major concerns of the Europeans, the animal health issue and its implications for human health and the impact on milk consumption of the introduction of the drug. Nor did JECFA's recommendations take into consideration the desire of the Europeans and the North American critics to apply a more stringent standard of safety to a product with no therapeutic benefit. In the United States the decision to approve rbGH involved more than considerations of safety; it involved the tacit approval of a particular agricultural system. Although JECFA's decision was made with regard to human safety, its recommendation assumed the appropriate use of the drug, and appropriate use involved management and agricultural systems that, in its recent concerns for animal welfare, the EU was beginning to reconsider. Because JECFA was specifically focused on human rather than animal health, its recommendations could not give adequate consideration to an alternative position on questions of animal welfare and dairy practice. The appeal of the United States and the industry to science does not allow for an alternative conception of assumptions about agricultural practices that are nonscientific but were nevertheless crucial to the decision in the United States.

Although a Codex ruling would not have forced the EU to approve rbGH and would not have changed its practices with regard to U.S. imports, a moratorium on the use of rbGH might have been increasingly difficult to maintain in the face of a positive Codex decision. The lack of consensus on the issue speaks to the importance of the domestic context in determining the policy implications of scientific evidence.

CONCLUSION

The rbGH case was a complex one. There was a surprising degree of consensus regarding the scientific evidence concerning human health. Even scientists within Health Canada who had raised concerns about the regulatory handling of the drug agreed that it was unlikely that it would have adverse effects on human health. Where they differed from the proponents of the drug and from the expert panel that reviewed the decision was in their perception of the role of the regulatory body. Because the standard data package had not been submitted, they argued that the appropriate regulatory standards had not been applied in the rbGH case and that those standards were warranted because anomalous results contradicted assumptions about the potential absorption of the drug that had guided the initial regulatory assessment. Expert review panels in both the United States and Canada, on the other hand, argued that long-term testing of rbGH would violate existing regulatory standards, given what was known about the drug. Only examination of the product after its release could determine absolute safety. For the critics of the product, this was precisely the problem.

The case highlights regulatory scientists' dependence on existing knowledge and the difficulty of questioning such knowledge within existing regulatory structures and understandings. The oversights in the regulatory process pointed out by the Canadian scientists may not undermine the decision, but they do highlight the fact that given the various pressures in the system, regulatory scientists are unlikely to question – and perhaps even to notice – anomalies that contradict their expectations.

Scientists who remained outside the debate also agreed that adverse effects were unlikely, but they thought that existing methodologies would not be able to prove this.

With regard to the animal health issue, again, scientists were largely agreed that the use of rbGH could increase common ailments. However, they disagreed on the implications of this finding for dairy farmers and for animal welfare and human health. In Canada and in Europe it was eventually concluded that the drug could be rejected on animal welfare grounds. In the United States, the animal welfare issue was not considered. The human health impact was considered in relation to milk contamination with antibiotics, but it was concluded that such a risk could be managed within the existing system.

Canadian scientists, however, argued that evidence of animal illness raised questions about human health, not animal welfare. However, there was no capacity within the regulatory system to explore this concern, nor could a detailed examination of the issue of antibiotic resistance be undertaken.

Different perceptions of safety issues and the context into which products are introduced at the national level have results in different decisions in the United States, Canada, and Europe, and they indicate the difficulty of imposing international standards. The Codex Alimentarius Commission has allowed the possibility of considering "other legitimate factors" at the international level; however, these factors are not yet clear, nor is it clear how we might arrive at a consensus regarding their specification.

6 Conclusion

THE INTERPRETATION OF THE EVIDENCE

In assessing the scientific evidence regarding the safety and effectiveness of rbGH, scientists within the company, the FDA, Health Canada, and the universities operated on the basis of existing scientific knowledge when assessing both animal and human health. The approval of the drug in the United States can best be explained as the result of a regulatory system in which the studies requested and the interpretation of the data were derived from a conventional scientific framework. I am using the term "conventional" here in Thomas Kuhn's sense to refer to science that has been accepted by an established scientific community – he also calls it "normal" science. Kuhn attributes the success of the sciences in solving puzzles about the natural world to the existence of such frameworks, or paradigms, which enable the existence of certain kinds of entities and relationships to be taken for granted and indicate which areas need still to be explored. Paradigms are necessary not only for the progression of scientific knowledge but for the formation of any coherent thought about the natural world. It is only within a particular framework that scientists are able to determine the relevance of their data and to distinguish anomalous instances from overall patterns.

In this case, conventional science included the literature on human studies conducted with natural bGH in the 1950s and, since rbGH was a protein hormone, the literature on protein digestion and absorption. Conventional science constructed reviewers' expectations about the likely result of safety studies, enabled them to decide what kinds of studies should therefore be requested, and influenced their interpretation of the data. Initially, the reviewers' consideration of the human health implications of hormone use focused on the biological activity of rbGH. The bGH studies of the 1950s had demonstrated that the pituitary growth hormone was not active in humans; it was therefore expected that the recombinant version would not be biologically active either. Experiments on rats were undertaken to test this hypothesis, and the data confirmed reviewers' expectations. The decision to allow milk from investigational herds into the food supply was based on this data.

Further examination of the literature suggested that another protein, insulin-like growth factor I (IGF-I), mediated the effects of growth hormone. Reviewers' initial expectations about the effects of IGF-I were formed by their understanding of conventional scientific knowledge about the substance itself and the class of compounds it belonged to (proteins). Based on scientists' knowledge of conventional science, however, it was not expected that the ingestion of IGF-I in milk from treated cows would have any effect on human health, and IGF-I studies were regarded by scientists as unnecessary. But controversy about the drug led the FDA to request studies in spite of the reviewers' convictions about the unlikelihood of risk. The study parameters were also determined by reviewers' expectations. The length of the IGF-I studies was set according to the length of time one would normally expect to see an effect from administration of a protein.

Generally, the resulting data confirmed scientists' expectations, although anomalous results were reported. Anomalies were apparent only in contrast to an expected pattern that had already been established by earlier studies. In instances where such anomalies did appear, they were also viewed through the conventional paradigm, in order to decide what effects could reliably be attributed to the drug itself. Regulators distinguished between *statistical* and *biological* significance. That is, they did not assume that a statistically significant difference between the treated and control groups could automatically be attributed the drug but considered whether the changes were consistent and whether they made sense in terms of existing scientific knowledge about its potential effects. So, for example, when female rats displayed effects not shown in male rats, these effects were not attributed to the drug, since nothing in the literature indicated that the effects of IGF-I were sex-specific.

Conventional science formed the knowledge base for academic and corporate, as well as for regulatory scientists. However, the pressures under which regulatory scientists operate tend to make them more reliant on conventional knowledge than their counterparts, whose work does not perform a specific public policy function. In order to determine which studies were most appropriate and in order to interpret the data from those studies, regulatory scientists employed the assumptions and beliefs that were prevalent in their own training and accepted by the established community of which they were a part. Regulatory scientists relied on conventional science, particularly as it was expressed in the literature, because their mandate precludes basic research and requires that they reach a conclusion that can be justified to the drug sponsors within a given period of time. This conclusion, as Salter (1988) has pointed out, will also eventually be subject to public scrutiny, but the immediate recipient is the sponsoring company.

There are several dimensions to the time limitation. Scientists in both the United States and Canada have 180 days in which to review the data package submitted to them by the drug sponsor and to decide whether to approve the drug, deny approval, or request further information. The time limitation does not compel a reviewer to approve a product he or she believes to be "unsafe." However, reviewers were conscious of the importance of timeliness for their own career prospects (timeliness was one of the criteria by which they were assessed at yearly performance reviews) and for the sponsoring company. They were aware that a request for further studies imposed costs on the company, both because of the resources expended on the study itself and because the additional time would delay market release. When competition between companies was fierce – as in the case of rbGH – the cost of delay was particularly high.

Reviewers would not violate their own definitions of safety in order to approve a product. However, their definition of safety was bounded by notions of what they could reasonably ask a company to perform, which depended, in turn, on the distinction between conventional and unconventional science. Human health evaluators regarded it as reasonable to request further information, in spite of the financial burden this implied, provided that there were good scientific grounds for such a request. Given that scientists' expectations and their interpretation of anomalous data were largely determined by conventional science, I would contend that "reasonable" requests must therefore remain within the boundaries of normal science, and the investigation of anomalies that would not be regarded as problematic within the established framework would not necessarily be regarded as reasonable and could therefore be regarded as placing an unnecessary burden on drug companies.

The system in which companies, academics, and regulators operate is not merely external to them, exerting its pressures from without, but becomes incorporated into their conceptions of reasonableness and timeliness – and therefore of safety. This is not to suggest that reviewers are automatons, uniformly reproducing their mandate. Individuals have a high degree of latitude in deciding whether a product is "safe"; in both Canada and the United States the decision regarding crucial elements of the human health review was made by a single reviewer and supported by his or her superiors. In forming their decisions, however, scientists work with an internalized understanding of their role within the institution in which they operate and the consequences of their actions for those affected by them.

The reliance on conventional knowledge was problematic for a product as controversial as rbGH. Critics believed that the anomalies could not be explained by the existing paradigm. In the absence of any other forum for questioning the appropriateness of agricultural biotechnology in general and rbGH in particular, critics seized on reports from scientists outside the debate whose conclusions represented a challenge to the conclusions of the academic, corporate, and regulatory scientists who contributed to the decision-making process about the drug. Not all scientists approved of the use of their work by the opponents of the drug. The conclusions these scientists reached, however, would seem to confirm some of Kuhn's propositions regarding the difference between conventional and unconventional science. The work that presented a challenge to conventional views about protein breakdown was conducted by researchers who were not specialists in the field and who had investigated it as a consequence of other research they were pursuing, rather than in their normal line of work or in response to the rbGH debate.

Viewed from the perspective of this research, the anomalies in the studies submitted to the FDA and Health Canada warranted further investigation. Critics wanted the FDA to review its human health decision in the light of more recent research. Although reviewers recognized that the critics' concerns could not be ruled out completely, within the understandings of normal science the critics' inferences were nonsensical and could not justify the expenditure of further resources to explore what they believed had been adequately investigated within the paradigm. Once a conclusion had been reached with which the reviewers were satisfied, their role and their own sense of self required that they stand by that decision. According to their understandings of existing science, the likelihood of a human health risk was virtually nonexistent. From the critics' perspective, however, even a low risk was not acceptable when the drug itself was perceived as offering few benefits against which the attendant risks could be weighed.

In Canada, the human safety issue created controversy within the regulatory agency as well as outside it. The chief of the Human Health Division of the Bureau of Veterinary Drugs, Dr Man-Sen Yong, decided that the use of rbGH did not represent a risk to human health. Reviewers within the bureau dissented from Dr Yong's interpretation and objected to a decision having been made unilaterally. The human health decision in Canada was interlinked with questions about the bureau's decision-making processes and degree of autonomy from corporate pressure. The dispute about the interpretation of evidence occurred in an environment in which reviewers alleged that their recommendations regarding other animal drugs had been overruled by managers who had been pressured by drug sponsors.

The critics also raised questions that had not been considered within the regulatory process and that were linked to broader questions about the consequences for human health of perceived health risks and the potential to aggravate latent risks in the food system. For the critics, the health and safety questions could not easily be separated from concerns about economic risks. Farmers feared that the perception of a human health risk would cause milk consumption to drop, further threatening livelihoods that were at risk due to increased production. Consumer and food policy groups noted that there were not only economic but also health consequences of reduced milk consumption. These kinds of secondary questions, however, could not be addressed through existing regulatory mechanisms, nor did individuals working with the relevant institutions think that they needed to be addressed.

The FDA's human health decision was based not only on conventional scientific knowledge but on what I have called contextual knowledge – a term adapted from Helen Longino's definition of contextual values, judgments about the way things are or ought to be. In order to decide whether differences in IGF-I levels between treated and untreated cows in investigational trials meant that treatment posed a risk to human health, reviewers considered such factors as the level of the growth factor in human saliva and breast milk and in pasteurized milk and infant formula. The scope of contextual knowledge was expanded by the GAO's investigations, which pushed the FDA to debate the use of antibiotics on dairy farms and the effectiveness of the institutions in place for ensuring that these practices did not lead to the contamination of the milk supply with antibiotic residues. After the GAO had advised that the product should not be approved until this issue was resolved, the FDA opened the issue to its advisory committee for consideration. At a meeting the FDA contended that public health would be protected by the existing state and federal monitoring system, which had been strengthened by recent reforms. These reforms had been instituted, however, as a result of previous GAO criticism of the system's failure to test for a wide enough range of antibiotics. A more recent GAO report suggested that although the FDA had expanded testing, the agency had failed to keep up with antibiotic use, and there were still many drugs that were not being detected. In the judgment of company and FDA scientists, the institutions currently in place could protect public health.

With regard to animal health, the data were more ambiguous, and the animal health decision cannot be detached from understandings about the nature of the existing system of dairy production. Contextual knowledge, therefore, played an important role in the animal health decision. The data from the animal health trials was difficult to interpret because rbGH did not induce any specific toxicological effects but resulted in an increase in common animal health problems. The extent of the increase could not be clearly determined by studies specifically designed to test safety, which involved a small number of animals, and it was therefore decided that health variables would also be measured during the efficacy trials, which involved larger numbers of animals. But because even these trials were not sufficiently large to base a conclusion on individual trials, the data were pooled. However, the incidence of health problems varied from herd to herd, complicating data pooling.

Under these conditions, the animal health data indicated that an increase in the incidence of mastitis and of various reproductive problems was associated with rbGH use. Reviewers had to decide, therefore, not only what level of incidence of disease was associated with rbGH use but also what level was acceptable. As with the human health data, reviewers tried to determine the *biological* significance of *statistically* significant increases in these variables. In order to draw conclusions regarding biological significance, reviewers compared the incidence of disease in treated animals not only with the control group but also with the general population under a range of conditions. They also considered whether the increase was *subtle* or *catastrophic*. A subtle increase was acceptable, whereas a catastrophic increase clearly was not. They also assessed whether the increase was *manageable*. "Manageability" had slightly different meanings in human than in animal health. The human health impact was partly determined by an assessment of the capacity of existing institutions to monitor and prevent antibiotic contamination of milk. With regard to animal health, an increase in the incidence of disease could be regarded as "manageable" if the *existing* situation was assumed to be acceptable and if the increase did not exceed a certain threshold amount. These calculations were contextual in the sense that they relied on an assessment of the acceptability of existing dairy practices, farmers' abilities to cope with current and potentially increased disease levels, and the point at which such increases would no longer be "manageable." These judgments could be contested, but not answered definitively, on the basis of strictly "scientific" evidence. Although corporate and regulatory scientists disagreed about the cause and extent of animal health problems, they agreed that an increase in the incidence of the disease would be "manageable." This conclusion was based on the premise that existing levels of disease were acceptable.

Existing dairy practice was reflected in both the trial protocols and the analysis of the data. Guidelines were developed by the FDA and its consultants for recording mastitis data after trials indicated that the disease was problematic. These guidelines attempted to balance the requirement for clear scientific information with the need for the trials to replicate real-world conditions, which involved the use of antibiotics to treat mastitis and other infectious disease and of hormonal treatment for the regulation of reproduction. The use of these medications was therefore permitted in the trials.

When interpreting the data, the FDA deliberated about what level of increase in disease farmers could manage. Their deliberations were based on their own experience of dairy farming and their training, as well as on a reading of the literature, attendance at conferences, and their interaction with the companies. Ideas of manageability were based on comparisons with notions about what a "successful" farmer was currently managing. Although the reviewers I interviewed at the FDA did not say so, their deliberations about the manageability of the problem must, in my view, have entailed an assessment of the economic impact of the drug. Mastitis is the most costly disease for the farmer because milk from mastitic cows must be discarded. Reproductive disorders also involve an economic cost to the farmer. I would take the reference to a "successful" farmer to mean *economically* successful, and the calculations about drug safety therefore involved considerations about the extent to which farmers would be affected economically if they adopted the product. An academic consultant claimed that the FDA assessed the impact of mastitis by comparing the expected increase in milk production with expected milk loss due to the disease and concluded that the productivity increase would outweigh the loss. I have not had this claim confirmed by the FDA.

Critics were skeptical about the concept of manageability, which appeared to them as an arbitrary distinction. At the interviews I conducted, "manageability" initially seemed to be a difficult concept to articulate because it was based on reviewers' tacit knowledge from their own experience and training. However, the FDA did come to a conclusion about what level of mastitis was acceptable and informed the company of what the "threshold" was. This limit was not expressed to the critics, however, who remained confused about its basis.

The critics also contested the statistical methods used by the FDA and the company to reach their conclusions about the extent of the increase of mastitis. The differences in methodology are less important as a cause of controversy than the background assumptions motivating them. From the FDA's perspective, the increase, when compared with existing conditions, was acceptable. The critics disputed this conclusion and the comparative method underpinning it, but they also disputed the assumption that existing conditions were acceptable. The introduction of rbGH occurred during a period of rural crisis, in which the number of farm closures, loss of income, and social breakdown in rural areas had reached proportions not seen since the 1930s, when forms of income support and import protection were first introduced to compensate for the devastating effects of the Great Depression. As that system began to buckle under the strain of farm debt and overproduction in the 1970s, farmers questioned the application of technology that had contributed to vastly increased productivity, which, in the context of farm support, had led to the overproduction crisis.

The contrast with the Canadian example serves to highlight the potential for different interpretations of the data and the extent to which judgments about acceptability depended on contextual knowledge. In the Canadian case, it was believed that the data raised serious animal health concerns. It is possible that in the initial stages of the review, the FDA shared similar concerns. However, a former Health Canada scientist did not agree with the FDA's conclusions and operated with very different conceptions of what constituted a subtle as opposed to a catastrophic effect. The threshold of acceptability would appear to have been much lower in Canada.

Monsanto and the university researchers who acted as principal investigators on product trials at land-grant universities in the United States and agricultural colleges in Canada had also concluded that the drug was safe, but the reasons for their conclusion differed from the FDA's. University and Monsanto scientists attributed the increase in mastitis associated with the drug to increased production levels and to the higher pretreatment incidence in treated animals compared to controls. Although some Canadian scientists were concerned that the attribution of mastitis to high levels of production rather than to the drug had not been sufficiently well established, they did accept the relevance of the relationship. The FDA, however, did not.

The FDA did consult with the company during the review process, however, and when deciding on the wording of the accompanying product label. It is difficult to determine what effect these consultations had on the FDA's conclusions. The company was able to obtain changes to the protocol that appear to be relatively minor but that did enable the company to provide evidence for its contentions about the incidence of mastitis in the treatment groups. By ensuring that the protocols measured the incidence of disease before treatment, the company was able to argue that this was an explanatory factor.

Although there were differences between the company's and the FDA's interpretation of the animal health data, ultimately both groups of scientists agreed that the risk involved was acceptable. This conclusion was rejected by groups that regarded the existing situation as unacceptable. The animal health decision and the reaction against rbGH needs to be placed in the context of the 1980s farm crisis. The literature on the political economy of agriculture has explained the crisis as the result of a crisis in the industrial model of production into which agriculture has been integrated. The model consisted in a system of mass production and consumption and relied on a complementary system of regulation. In the post-World War II era, the use of industrial inputs in agriculture, which had commenced in the 1920s and 1930s, expanded further and was accompanied by the transformation of farm production into the production of industrial inputs for processing by food conglomerates. These two trends meant that agriculture was no longer a discrete sector but had become a subordinate part of the industrial production system. These trends, however, had initially been facilitated by nationally based farm-support programs that had been introduced in the 1930s to protect farmers from the vagaries of the international marketplace and to reduce the disparity between urban and rural incomes.

The deployment of state price-support programs in the United States and the adoption by other Western countries of similar programs modelled on the U.S. system constituted one aspect of what Harriet Friedmann has termed the "food regime," an implicit framework of rules governing the regulation of food production on a global scale. Although the policies aimed at domestic protection, they also facilitated the integration of food production across national lines. This combination of domestic protection and transnational production characterized the "surplus regime," in which overproduction of wheat, soybeans, and cheap oils was disposed of through food aid programs and export of feed grain for cattle as meat consumption expanded. In the early 1970s, however, this regime collapsed with the export of massive quantities of wheat to the Soviet Union (the surplus regime had previously excluded the socialist bloc), the breakdown of the Bretton Woods monetary and trade regimes, and the oil price hike. Although the transformation of agriculture into a segment of global, industrial production was predicated on government support programs, transnational production has since become decoupled from domestic regulation. But new systems of regulation are gradually emerging. For the first time since negotiations began, agriculture was included in the Uruguay Round of the GATT negotiations, with the United States aiming to reduce agricultural protection schemes worldwide.

The introduction of rbGH and the resistance to it needs to be placed in the context of the breakdown of the surplus regime, with its component systems of national regulation and the extension of the transnational restructuring of food production and diets, coordinated by transnational agri-food corporations rather than by national governments.

Monsanto, an agribusiness corporation that had created a global market for its herbicides, perceived that biotech would have a dramatic impact on the global restructuring of agriculture. The company persisted with drug development in spite of the resistance because of its belief that its future profitability depended on investment in biotechnology and because rbGH was the first biotech product in the pipeline. As we have seen, the company's interest in developing some form of bGH dated back to the 1960s, when scientists had attempted to synthesize a chemical version of the hormone. The advent of recombinant technology enabled this project to be fulfilled. By the 1980s, however, the use of recombinant technology to produce this specific product was less significant than the creation of the technology itself. Monsanto chose rbGH because it was one of the first molecules to become available; had the technology initially been applied to another molecule, Monsanto would have pursued another line of development. Monsanto, like other chemical firms, turned to biotechnology in order to offset declining profitability in chemicals. In the 1970s leading chemical, oil, and pharmaceutical firms turned to biotechnologies as part of a dual strategy to revitalize profits by diversifying production and extending demand for their agricultural chemicals. In Monsanto's case, agricultural biotechnology fulfilled both goals. As early as the 1960s, the company had sought ways to extend the market for its top-selling herbicide, Roundup, and to develop alternative products. In the 1980s, it used plant biotechnology to create Roundup-resistant cotton and soybean seeds. In 1997 it went further and divested itself of its chemical enterprises, renamed

Solutia, in order to focus on the application of biotechnology, or "life sciences," in agriculture and pharmaceuticals.

The belief in the significance of the biotechnology revolution was the overwhelming force behind the development of rbGH. The company proceeded with the drug and resisted protest against it because of the strength of this belief. The commercialization of the drug required decisions at several stages and at each stage the company persisted with the product in spite of regulatory and political conflicts. The company made its own assessment of the safety and efficacy of the product before submitting an application for regulatory approval. This assessment was based on existing scientific knowledge and the results of short-term trials. Decisions were also made based on assessments about the marketability and acceptability of the drug, but the company failed to anticipate certain regulatory requirements, as well as opposition from farm, consumer, and food policy groups. The initial route of administration of the product was based on marketing considerations. The marketing department believed that farmers would only accept a product that could be administered like antibiotics, but this route affected meat quality and hence was not acceptable to the FDA.

When Monsanto began developing rbGH, company scientists and managers believed that the product would appeal to farmers, because they assumed farmers shared the goal of increasing efficiency. Farmers, however, particularly in the main dairying states of Wisconsin and Vermont in the United States and in Canada, were far more ambivalent about the utility of efficiency in a period of record milk surpluses. In the United States, farmers questioned the wisdom of enhancing productivity, not only because they doubted the need for further milk increases but also because they had begun to question the appropriateness of earlier technological innovations that had resulted in vastly increased productivity and a corresponding reduction in farm numbers. Monsanto was surprised by the reaction, particularly from small farmers, since unlike earlier dairy technologies, rbGH did not require a large capital investment and therefore was equally available to holders of small and large farms alike. Company scientists believed that since the technology did not require a huge capital investment, it was "scale-neutral" and did not represent a threat to small farmers. Protest against its introduction, however, was based not on perceptions of its inaccessibility but its the net effects on the dairy surplus and, consequently, on dairy support programs.[1]

The company continued to hold that the drug could benefit all dairy producers and when the drug was approved, provided incentives that did not discriminate between large and small farms, in order to encourage the adoption of the drug among farmers with smaller holdings. Scientists and managers were surprised by the reaction to their message, rather than by the reaction to the product itself. The company believed that resistance could be overcome through the dissemination of information. Monsanto needed to convince its own employees, as well as those outside the organization, of the wisdom of introducing rbGH. It also needed to reach consumers to convince them of the safety of the product; their acceptance was important to its marketability, since without it, farmers may not have been willing to purchase it. In order to reach those outside the company, it enlisted the support of medical associations, who also engaged in a process of convincing their own members to back the association's position on rbGH and on biotechnology in general. Questions and uncertainties about biotechnology were expressed within the American Medical Association and the American Dietetic Association before the development of position statements on these issues. But by the time the statement had been produced, dissent had been quelled and the entire membership was prepared to back a position originally articulated by an individual or a small number of individuals within the group.

The universities were also involved in the debate. University scientists assumed a dual role – as evaluators and endorsers of the safety of the drug. They contributed to public policy not only through their work but through their active involvement in the debate about safety. Scientists with responsibilities in extension services as well as in teaching and research informed their constituents about the safety of rbGH, forming networks across disciplines in order to be able to refer questions to the appropriate expert. The extension role, particularly in the United States, became more important as the controversy surrounding the product intensified; indeed, it was determined by the extent of the controversy. This role fulfilled several functions for the companies. In 1991 the companies were forbidden from directly promoting the safety of the product, but the involvement of third parties in the debate meant that support for the product could still be articulated. Their involvement also diffused controversy and enabled the company to reach consumers more directly. The researchers, however, did not perceive anything problematic in their role but regarded it as a contribution to public education that enabled farmers, processors, and consumers to make an informed choice about the product. The scientists' perception reflects the diminution of public research to the extent that the distinction between "public" and "private" goods has been blurred.

In their extension role university scientists were expected to articulate their position on the safety of the drug, although their information was based mainly on their experience from trials conducted at their institutions, trials that may or may not have reflected the overall situation. Unlike the FDA or the companies the university scientists did not have the pooled data at their disposal, and the individual trials were generally conducted with too small a sample of animals to draw firm conclusions about product safety. After the product had been approved by the FDA in the United States, however, university scientists and Monsanto scientists published a coauthored paper that analyzed data pooled from fifteen trials in America and Europe. The scientists were also aware that the company data was proprietary and that companies were competing in the race to get the product to market first.

The relationship between the universities and the company was complex. Both parties gained from the relationship. University studies could be conducted more carefully than field studies and the data could be more accurately recorded. University studies also had greater legitimacy than those conducted at company sites. Companies were able to draw on the skills and resources of several different disciplines within a single institution and to economize by linking the research performed in each area. For example, food scientists could test the nutritional quality of milk from cows in herds being tested by animal scientists within the same institution. In return, the scientists were able to pursue their own research interests by piggybacking on the company studies. The company provided additional funding that paid for technicians and materials and thereby allowed university researchers to examine questions they were interested in. In Canada, such financing enabled scientists to apply for matching government funding.

Not only were professional associations and university scientists involved in the rbGH debate but the FDA itself was involved in responding to queries and answering critics. My analysis here relates closely to Salter's (1988) discussion of mandated science, that is, science that serves a public policy purpose. One of the pressures that mandated science experiences, in contradistinction to normal science, is the pressure to meet the requirement of public openness while simultaneously addressing the scientific community and maintaining the confidentiality of data. Only an idealized science, Salter contends, could fulfill these conflicting goals. In the rbGH case, conflicting pressures were experienced by scientists very early in the evaluation process, before they had evaluated the animal safety data. The companies and universities conducting contract research publicized the results of their studies very early in the review process, with the result that public alarm about the impact of the drug was raised early and the conclusions about human and animal health were anxiously anticipated. To allay public fears, the FDA took the unprecedented step of publishing a summary of the human health data in the respected journal *Science*. It also presented some data for review by the National Institutes of Health and opened its decisions regarding the potential indirect public health risks and product labelling to its advisory committees, another unprecedented move for an animal drug. The FDA also publicly endorsed the safety of the product, in an effort to defend its own credibility, as well as to ensure that time limitations were not exceeded. When inquiries from the public began to affect reviewers' capacities to complete their evaluation in a timely manner, they defended the human health decision publicly. The legitimation process in the case of rbGH, then, took place not merely through the act of regulation itself, but through the reviewers performing the dual role of evaluating the product and legimitizing their decision to the public.

The form of "progress" represented by agricultural biotechnology was highly contested during the investigational trials. The creation of the Coordinated Framework for the regulation of biotechnology in 1986 and the congressional hearings on the ethics of transgenic animal patents in 1987 put biotechnology back on the agenda of public debate. In public discussions, FDA officials tried to distinguish between food derived from rbST-treated cattle and "transgenic" products. Biotechnology policy was based on the distinction between "process" and "product," however. When discussing the Coordinated Framework, FDA scientists observed that the framework was not relevant to the rbGH issue because the product was not "biotech"; however, it was the framework itself that promulgated a definition of biotechnology that excluded rbGH from this category. Policymakers distinguished between the *process* and the *product*: the product could be regulated through the existing system; the process was not inherently harmful and did not, therefore, require separate legislative action (see Jasanoff 1995a). This distinction was ironic given the other discourses about the significance of biotechnology. On the one hand, the belief in the capacity of technology to revitalize the American economy motivated the changes in American patent and tax law and facilitated the commercialization of biotechnology. The economic implications of this new development were regarded as so profound that proponents warned that regulatory restrictions would kill investment and put the United States at a comparative *dis*advantage. On the other hand, the techniques themselves were not regarded as revolutionary but as a mere extension of traditional means of genetic selection (see Kessler et al. 1992).

The FDA evaluators attributed the start of the controversy to the actions of a few individuals who were motivated more by concerns about biotechnology and the over-production of milk than by the human health implications of the drug. They also believed that the symbolic value of the product inflated the sense of public danger. FDA officials are surely right in concluding that public anxiety is more likely to be aroused by a perceived danger from milk than from other products (witness the concern about rbGH compared to the potentially more damaging evidence regarding hormone use in beef cattle). Protest groups were well aware that the manufacturers' choice of biotechnology product did not help their case. However, the image of milk purity had been heavily promoted by the industry itself, and public alarm at the prospect of its adulteration was at least partly a reaction to past idealization.

However, although the agency spoke out about the portions of the data it had evaluated, much of it remained confidential. Confidentiality created problems, particularly in Vermont, where the authenticity of the data was challenged by rural advocacy groups and congressional representatives. Scientists outside the debate, whose results conflicted with those of the FDA, wished to see data that indicated how the experiments were performed. They may indeed have been performed to their satisfaction, but without a description of the methodology they were not prepared to accept the conclusions reached.

The difference between purely scientific debates and those centred around public policy issues was particularly apparent in the rbGH case. The language in which it was expressed emphasized the public policy conclusion rather than the anomalies in the data. Scientists outside the debate, on the other hand, tended to emphasize the experimental constraints under which their work was conducted and the difficulty of extrapolating from it. Finally, rbGH science, particularly in its estimation of the animal health data, could not be easily extricated from considerations of the impact of the drug.

Once a decision on the safety of the drug had been reached, the FDA also considered whether milk from treated cows should be labelled as such. Critics, on the other hand, wished consumers to be informed through mandatory labelling on dairy products from treated cows. They had hoped that if the drug was authorized for sale, those who opposed its use could effectively mobilize against it by voting with their wallets and refusing to buy the milk. The FDA was aware of its legal obligations and constraints with regard to labelling policy, however. Although the agency recognized consumers' right to know about the content of the food they were consuming, it also recognized that the right to know had not been broadly construed. Under existing guidelines, labelling could not be required unless consumers were deemed to have a material interest in the product label, that is, if the milk presented a human health risk or unless its organoleptic qualities (taste, smell, texture) differed from milk from untreated cows. The agency was aware that by going beyond these guidelines, it risked imposing a greater obligation on the company than was required by the statute and that such a labelling requirement could be regarded as false and misleading if it implied that the milk from treated cows was inferior to that from untreated animals. The FDA did recognize a consumer interest in product labelling by permitting voluntary labelling of milk from untreated animals, but in order to avoid the implication of inferior quality, the guidelines stipulated that any label noting that a product was rbGH-free must also state that there was no significant difference between milk from treated and untreated cows. The scope of labelling, therefore, was largely determined by the health decision, although critics had hoped that it would provide a way of resisting the decision that had been made.

Voluntary labelling created a number of problems. It was the farmer who was *not* using the product who had to demonstrate that his or her product was rbGH-free. Soon after the product was approved, Monsanto threatened two companies who were labelling their product as rbGH-free with legal action, although their label also included the "no significant difference" disclaimer. Conflict erupted between states when guidelines in one state, which outlawed any mention of rbGH on product labels, resulted in the disposal of products from another state that had instituted labelling guidelines. Voluntary guidelines were regarded as insufficient by states that had campaigned hardest against rbGH labelling laws, but their attempts to introduce mandatory labelling were overturned by the courts, who did not recognize the relevance of consumer interest. In Canada, labelling has been ruled out because of the cost of creating a dual-supply system to separate milk from treated and untreated cows.

CONCLUSION

This book has argued that the problems in the relationship between science and policy arise at the point at which judgments are made about the implications of the data. In the case of rbGH, although there was a consensus about what the data showed, interpretations varied between contexts, the contrast between Canada and the United States being the most obvious example of this.

A framework based on Longino's concept of contextual knowledge best explains the rbGH case. Although Longino argues that we make sense of empirical evidence in relation to the background assumptions we bring to it, she does not deny the relevance of the evidence or the importance of the scientific enterprise. In the rbGH case there was a surprising degree of consensus about the evidence, even among the critics. The hypothesis to which this evidence was related was ultimately, however, a social one. A conclusion about safety can only be informed by data, not determined by it, and it was this conclusion that was the source of dispute.

The conventional knowledge that informed the scientists' construction of guidelines and their interpretation of data can also be seen in Longino's terms as forming the basis for background assumptions. The extent to which scientists departed from this conventional knowledge depended on their institutional affiliation and, in some cases, their specific discipline. Although this did not necessarily determine their answers to questions of safety, it did influence answers to the various parts of the argument needed to reach conclusions about safety.

In the case of animal health and the potential human health consequences background assumptions were also related to the context into which the drug was to be introduced. Given that the animal safety debate was conducted in terms of questions of production and management, there is no way to disconnect it from contextual considerations.

The debate about the safety of rbGH/rbST ultimately relates to our expectations of the regulatory system and the agricultural system. At the time of the U.S. approval, the data did not indicate potential harm to humans. Furthermore, alternative hypotheses did not indicate a public health risk but suggested that certain premises on which human safety conclusions had been based warranted further exploration, but this was research beyond the scope of regulatory science in its existing form. When issues regarding the application of biotechnology have not been resolved socially, the demand for them to be resolved scientifically will further increase the pressure on regulatory bodies. The development and approval of rbGH in the United States was the outcome of a series of decisions at the regulatory, academic, and corporate levels that fostered the rise of the biotechnology industry and limited regulation of its products to a technical evaluation of health and safety under existing law. Concerns about social and economic impact of rbGH and about the long-term human health effects could not be identified within a framework that excluded the former considerations and evaluated the latter in terms of the conventional scientific framework.

Although there was no evidence of human harm, there was evidence of adverse animal effects. In the United States these effects were regarded as manageable and therefore acceptable, but not in Canada. If we do consider the drug as a management or production tool that can increase efficiency in conjunction with the application of other skills, the question of its acceptability comes down to the kind of production system we wish to have. The related human health questions can not be extricated from this question. The U.S. conclusions regarding this issue were based on the unlikelihood of animal drug residues in milk representing a public health risk. Other questions regarding antibiotic resistance and food-borne illness could not be answered within this framework, and rBST itself would be a factor here only in a broader set of practices. Compared, for example, with the use of antibiotics in animal feed at subtherapeutic doses or in treatment of animals at unapproved doses, the increase in medication associated with the use of rbGH is not regarded as problematic. As those practices themselves become subject to further examination, particularly in the wake of bovine spongiform encephalopathy (mad cow disease), spreading antibiotic resistance, and outbreaks of food-borne illness in locations across North America, the demands on regulatory science, as well as on other forms of science, will grow, unless we also develop alternative methods of considering these questions.

Notes

CHAPTER ONE

1 Even the name has been controversial. The term "growth hormone" has been used in many scientific reports about the hormone. However, proponents prefer to call it "somatotropin," which eliminates the word "hormone." The Europeans have argued that the correct term should be "somatotrop*h*in," since trophic factors affect growth, whereas *tropic* factors affect movement. The debate reflects the battle over meaning and interpretation that is critical to the controversy. I use the terms rbGH and rbST interchangeably, but generally I use the term rbGH. Monsanto's product formulation, sometribove, is called Posilac in the United States and Nutrilac in Canada.
2 Initial estimates were much higher – later reports indicated that this range was more likely.

CHAPTER TWO

1 Before 1980, federal agencies followed twenty-six different patent policies (OTA 1990, 54).
2 Joint research ventures were not regarded as a violation of anti-trust law *per se* but judged according to "reasonableness" and "relevant factors affecting competition" (Slaughter and Rhoades 1996, 320).
3 Monsanto describes itself as a "95-year-old startup company." (1997a, 2).
4 By 1995, however, Monsanto's investment appeared to have paid off. The company reported record net income of $739 million and earnings

per share of $6.36, providing shareowners with a 79 percent return compared to the Standard and Poor's average of 37 percent (Monsanto 1996a, inside cover; Shapiro 1996, 1). It had been restructured into thirteen business units: five agricultural products units, five chemicals units, the Searle pharmaceuticals unit, and two food-additives units, including the Nutrasweet company, producer of the artificial sweetener aspartame (Monsanto 1995, 13).

5 Not all industry members agree with this position, however. Another major chemical firm, DuPont, has come under pressure to follow Monsanto's strategy and split its chemical operations from its life-sciences operations, but the company intends to maintain its current structure. DuPont's chief technology officer, Joseph A. Miller Jr., has commented, "Why not use molecular biology and genetics to make plastics and fibers?" (Deutsch 1998, C3)

6 Monsanto had already begun marketing two products engineered to express genetic characteristics from other organisms. In association with Asgrow Seed, Monsanto began marketing herbicide-resistant soybeans. The seeds had been engineered to express resistance to Monsanto's glyphosate herbicide, Roundup, thereby enabling farmers to use the herbicide directly on the crops without damaging them. The farmer must pay a $5 for each bag of the soybeans, use Roundup, allow inspection by Monsanto officials, and agree not to supply the seed to other farmers. The company had also developed insect-resistant plants. Genes from the *Bacillus thuringiensis*, or *Bt*, bacteria, which code for proteins that can kill insects, had been inserted in Monsanto's NewLeaf potato in order to make the potato resistant to the Colorado potato beetle. Monsanto had also commenced trials in the Mississippi Delta with cotton seeds inserted with the *Bt* gene, known as *Bollgard* cotton (Feder 1996, F1).

7 Monsanto had attempted a different merger in June 1998, a $33.5 billion deal with American Home Products Corporation, a company that, according to the *Wall Street Journal*, "sells everything from Chap Stick and Dimetapp to Preparation H." The merger would have enabled Monsanto "to move genetically engineered products ... into supermarkets and drugstores" (Kilman 1998, B10). American Home Products had bought into the pesticide business in 1994 with the acquisition of American Cyanamid and now had a $2 billion market in pesticides. Monsanto's pesticide business had been a major competitor. However, the deal was not completed; it was "doomed by personality conflicts," according to the Wall Street Journal (Kilman and Burton 1999, A10).

8 This decision was made before the passage of the 1986 Drug Export Amendments Act, which would have allowed the company to export the product to Europe before its U.S. authorization if European approval was granted first.

9 The United States was not alone in this. In 1987 the European Community spent $3.7 billion to dispose of more than 1.3 million tons of surplus butter, which cost more than $1 billion a year to store ("Europe to Cut Butter Stocks").

10 Although data from Canadian sites can be submitted to the FDA and although U.S. data are accepted at Health Canada, because Monsanto did not proceed with the formulation tested in Canada, information about these trials was not released in the FDA's Freedom of Information (FOI) Summary.

11 The first Morrill Act of 1862 established land-grant colleges in each U.S. state and territory and in the District of Columbia. The second Morrill Act of 1890 mandated the establishment of colleges for African-Americans located in the Southern states (National Research Council 1996, 1). The colleges are known as land-grant institutions because they were funded from the sale of land granted to the state by the federal government under the Morrill Act. Every state remaining in the Union was given a grant of 30,000 acres of public land for every member of its congressional delegation. Proceeds from the sale of this land were used to establish colleges in engineering, agriculture, and military science (Roberts 2001).

12 The Hatch Act of 1887 mandated the creation of State Agricultural Experiment Stations for the conduct of research in cooperation with the colleges of agriculture; the Smith-Level Act of 1914 was intended to transfer the results of this research to the local population (Hightower 1973, 1n2).

13 Monsanto was later criticized by the inspector-general of the Department of Health and Human Services for continuing promotional activities after the FDA warning (see "Monsanto Assailed," 1994; Office of the Inspector General 1991).

14 Jack Doyle has noted that approximately 57 percent of CAST's operating budget comes from two hundred agribusiness corporations and trade associations (1985, 368).

CHAPTER THREE

1 Jasanoff has noted that the capacity of the courts to affect policy has been circumscribed in the case of biotechnology by the 1986 Coordinated Framework for the Regulation of Biotechnology (1995b, 157). In the 1980s the courts also followed a new doctrine with regard to their assessment of agencies' regulatory decisions – "the mere existence of scientific uncertainty did not justify the regulation of trivial risks." In *Monsanto v. Kennedy*, 1979, the DC circuit judged that although the governing statute granted agencies authority to regulate even trivial risks, the agency

should not conform to this standard if the risk was insignificant (82).

2 The council was comprised of the commissioner of the FDA, the director of the NIH, the assistant secretary of agriculture for marketing and inspection services, the assistant secretary of Agriculture for Science and Education, the assistant administrator of the EPA for pesticides and toxic substances, the assistant administrator of the EPA for research and development, and the assistant director, biological, behavioral, and social sciences, NSF (Kingsbury 1986, 50).

3 These characteristics can be divided into two types. *Agronomic* characteristics affect plant yield – these include characteristics such as disease, pesticide, or herbicide resistance. *Quality* characteristics affect the processing, preservation, nutrition, and flavour of the product (FDA 1992, 22985).

4 I will use the term FDA to refer both to the agency and its veterinary medical division, the Center for Veterinary Medicine.

5 The plaintiffs included the International Dairy Foods Association, the Milk Industry Foundation, the International Ice Cream Association, the National Cheese Institute, the Grocery Manufacturers of America Inc. and the National Food Processors Association (Centner and Lathrop 1997, 540n).

6 In *Stauber v. Shalala* (895 F. Supp. 1178 W.D. Wis. 1995), the court held that the FDA's decision to approve a drug "may be set aside upon judicial review only if the agency's determination is arbitrary and capricious, abuse of discretion or otherwise not in accordance with law." The court also noted that "The arbitrary and capricious standard is highly deferential; even if a reviewing court disagrees with an agency's action, the court must uphold the action if the agency considered all relevant factors and the court can discern a rational basis for the agency's choice."

CHAPTER FOUR

1 Article 1709, paragraph 10 outlines the conditions under which a compulsory license may be granted. First, a license must have been sought from the patent holder on reasonable commercial terms. Second, the license must be granted predominantly to supply the domestic market. Third, if the circumstances that meant that licensing was required change, the license is to be terminated, and the patent holder must be paid adequate remuneration. Fourth, a license should not be granted to permit the exploitation of another patent (Food and Drug Law Group 1994, 335).

2 Under Bill C-91, this period was thirty months. Recent changes to the regulations have reduced this to twenty-four months (Manley 1997).

3 Protection for brand-name pharmaceuticals was further extended in the year 2000, when the WTO ruled that Canada must provide twenty years

of patent protection to patents issued before 1989, which had previously been given only seventeen-year protection. The Canadian government had argued that TRIPS was not retroactive and should not apply to patents issued before TRIPS entered into force in 1995 (Canada NewsWire 2000, 1).

4 The application to patent the oncomouse was rejected by the Canadian Intellectual Property Office (CIPO) in 1993. This decision was upheld by the commissioner of patents in 1995 and by the Federal Court Trial Division, which stated that "a complex life form does not fit within the current parameters of the Patent Act" ("Health Monitor" 1998a, 42). However, Harvard University, the owner of the U.S. and European patent, appealed the decision to the Federal Court of Appeal, which recognized that the mouse qualified as a "composition of matter" under the Patent Act. In the ruling, it was noted that there was no provision for a distinction between lower and higher life forms in the legislation and that Parliament was the proper forum for excluding certain types of subject matter from the legislation (Gravelle and Wong 2000, 2). However, the commissioner of patents will appeal the decision (Foster 2000).

5 The MRC has been replaced by the Canadian Institutes of Health Research (CIHR), which implements a multidisciplinary approach to health research facilitated thorough thirteen "virtual" institutes (CIHR, 2000).

6 A recent study, however, has advocated the adoption of the U.S. model. It has recommended that universities should adopt "innovation" as their fourth mission, after teaching, research, and community service, that researchers should be compelled to disclose potentially marketable IP to the institution with which they are affiliated, and that the institution, in turn, should be responsible for forwarding this information to the federal government (Expert Panel on the Commercialization of Research 1999, 4).

7 In 1998 regulators from the U.S. counterpart of the CFIA, the Department of Agriculture, Animal and Plant Health Inspection Service (APHIS) met with representatives from the CFIA, Health Canada, and Industry Canada to discuss the possibilities for harmonizing the regulatory process in the two countries that may lead "to mutual acceptance of assessments in the future" (CFIA 1998, 1).

8 CEPA was passed on 30 June 1988. It had replaced or subsumed several earlier acts intended to protect the environment. Section 139 of the act stipulated that a parliamentary committee would review its provisions within five years (ix). This task was undertaken by the Standing Committee on the Environment and Sustainable Development.

9 With regard to recommendation 4, the council argued that the agriculture committee had called for changes in the labelling system so that con-

sumers and regulators could determine whether imported products contained milk from treated cows, whereas the government's response indicated that the status quo was fine and ignored the fact that mandatory labelling legislation had been passed by some U.S. states. On recommendation 5, the TFPC pointed out that there was no evidence that the government intended to make any changes to the review process. It argued that recommendation 7 had not even been "partly" endorsed. An environmental assessment should be carried out for rbGH because "[a]ny sound environmental assessment is rooted in a belief that actions have a whole series of interconnected reactions, and that only a comprehensive determination of these relationships can lead to an understanding of the implications." The interactions in the rbGH case would include manure and manure management, feed demands, pesticide and fertilizer use, regional concentrations of dairy production, and land-use patterns. It pointed out that socioeconomic criteria had been used by the European Community to reject rbGH. Finally, the TFPC recommended that the composition of the task force must be changed so that at least half its members had concerns about rbGH licensing, that its mandate should not be restricted to a review of existing documents but should allow organizations to submit briefs and appear before it, that the government should take action to ensure that the moratorium was respected by importers, and that no NOC should be issued before the task force reported.

10 The task force was comprised of Ray Mowling from Monsanto and Terry Clark from Eli Lilly, Brian Morrissey from Agriculture and Agri-Food, David Head from Industry Canada, Dairy Farmers of Canada president Peter Oosterhoff, Dale Tulloch from the National Dairy Council of Canada, and Ruth Jackson from the Consumers' Association of Canada.

11 The grievance, regarding management interference with the veterinary drug evaluation process, had been filed with the department in 1996. The group grievance was dismissed by the associate deputy minister, Alan Nymark, as the outcome of interpersonal, rather than scientific, conflict. In response, the scientists, acting through their union, the Professional Institute of the Public Service of Canada (PIPS) took their grievance to the Public Service Staff Review Board (PSSRB) and also wrote to the prime minister requesting an inquiry into the drug evaluation practices at the Bureau of Veterinary Drugs. Their grievance was dismissed by the PSSRB. In 1998 Dr Shiv Chopra and Dr Margaret Haydon were interviewed on CTV's news program Canada AM, and were subsequently reprimanded by the director of the Bureau of Veterinary Drugs, André Lachance. They took a grievance against the letters of reprimand to the associate deputy minister, Alan Nymark, who denied the grievance. Subsequently, the case went before the Federal Court of Canada for judicial review. The court ruled in favour of the grievors, referring the matter back to the associate

deputy minister. In her judgment, Justice Tremblay-Lamer ruled that a public servant's common law duty of loyalty to his or her employer did not prevent him or her from disclosing to the public matters that were in the public interest (Federal Court of Canada 2000).

12 At the hearings Dr Margaret Haydon noted that in all the studies she examined for Revalor-H, there was a noticeable decrease in thymus weight of calves receiving the hormone. Because the thymus is important in the maturation of the immune system, she was concerned that the drug might compromise their immune response (Senate 1999c, statement by Margaret Haydon). Dr Chopra said that when he had supervised Dr Haydon's review of the product, a representative from the company had asked that she be removed from the review because she was too slow and "nitpicking" (statement by Shiv Chopra). In a memo to the director general of the Food Directorate, the acting chief of the BVD, Dr Donald Landry, noted that he had told Hoechst Canada that he would "make up for the rough time he's had with Revalor-H when we review his next submission" (Eggertson 1997b, A1). A representative from Hoechst confirmed that he had gone over the reviewer's head to ask both Dr Landry and Dr Paterson to give the drug review further consideration (A10).

13 The *Globe and Mail* reported that six scientists had written to the minister of health, Allan Rock, for assistance after their advice had been rejected by the head of the Bureau of Veterinary Drugs, Diane Kirkpatrick. Several drugs were in dispute, including a growth promoter in pigs, carbadox, which can leave carcinogenic residues in meat and enter the environment through the animal's waste. The scientists had recommended that the drug not be licensed in Canada, but Ms Kirkpatrick had requested additional information from the sponsor of the drug (Bueckert 2001, A4).

CHAPTER FIVE

1 A more detailed discussion of the debate can be found in chapter 5 of Mills (1999).

2 The Upjohn company's product was the exception to this; it produced a recombinant hormone with the same amino acid sequence. Monsanto's Posilac substituted an amino acid at the end of the sequence.

3 In the *Science* report, Juskevich and Guyer (1990) noted that "because of the general lack of information in the scientific literature regarding the oral activity of IGF-I, the CVM decided to obtain more information" (879).

4 Oral toxicity studies were requested from all companies that were pursuing approval at the time. Whether the companies conducted the studies depended on whether they planned to develop their formulation for regulatory approval (Juskevich 1998, personal correspondence).

5 However, Health Canada scientists noted that the method used to detect IGF-I in serum was not adequate (Senate 1999c, statement by Gerard Lambert).

6 The Toxicology Branch chief, Dr Judith Juskevich, did not review the studies, nor did she make the initial decision on the significance of the findings. She did, however, discuss the decision with the reviewer and agreed with the reviewer (Juskevich 1998, personal correspondence).

7 The provisions of the 1996 Animal Drug Availability Act (ADAA) were intended to prevent this outcome by making an agreement reached at the presubmission conference binding.

8 According to Hansen, "the only two risk categories of cows sampled are rabies-suspect cattle that are rabies negative, and cattle over two years of age that have been given protein supplements for a good part of their diet and have developed signs of neurological disease" (1993, 9).

9 The FDA noted that rbGH cows were treated more frequently with medication, including medication for mastitis. However, when the incidence of mastitis was examined on a per-milk basis, the effect of the drug was less than the effect of other factors such as season, parity (many times a cow has calved), stage of lactation, and herd-to-herd variation (FDA 1993d, section 6j). The FDA recorded data on the average duration of cases of mastitis in control and treated cows and concluded that the increased total days affected in the treated animals reflected the number of cases of the disease rather than its duration in treated animals (table 69). On the basis of data from individual trials and of pooled data, it was also concluded that treated cows were less likely to conceive and to calve successfully – that is, to carry a calf to, but not beyond, full term and for the calf to survive for more than seven days after birth without the mother having to be removed from the herd. The use of the drug was also associated with an increased rate of twinning and incidence of cystic ovaries. (When the drug was monitored for two years after its approval, no such incidence was found, however). FDA reviewers had not anticipated that the use of the drug would have any impact on lameness, but reports indicated that this problem needed to be examined. On the basis of studies specifically designed to investigate the issue, the agency concluded that the drug did not result in increased lameness but did result in increased foot disorders in multiparous cows and lacerations of the knee.

10 With regard to comments in the following section about the FDA review process, it should be noted that in order to avoid releasing proprietary information by specifying details of the rbGH case, an FDA scientist described the general process used by the scientists with the Center for Veterinary Medicine to advise drug sponsors on appropriate designs of studies to evaluate efficacy and animal safety.

11 This knowledge included an understanding of "dairy management, the

physiology of milk production, and what kind of changes might occur as a result of a production drug," according to an FDA scientist.

12 A further discussion of the role of land-grant universities can be found in chapter 2.

13 Extra-label use was permitted in the trials if use was not excessive, if records reflected that a veterinarian was directly involved in the decision to use the drug, and if the prescribed extra-label use was considered reasonable for the health condition recorded. When the pivotal studies for Posilac were being conducted, the FDA provided regulatory discretion to U.S. veterinarians prescribing extra-label drugs under those conditions. Furthermore, the Animal Medicinal Drug Use Clarification Act (AMDUCA) of 1994 affirmed the FDA's position and legalized such veterinary prescriptions.

14 According to one Monsanto scientist, the first Monsanto data submission was made in 1987.

15 Analysis began when data had been submitted. The FDA had designed an approach to summarizing and analyzing health data under categories of major body systems and subsystems. According to an FDA official, this approach allowed the centre a more organized approach to evaluating health effects of the investigational drug and appropriate labelling of the product if it was approved. Data from the acute toxicity study, the 1, 3, 5X animal-safety study, several efficacy studies, and additional pivotal studies formed the animal safety and efficacy package. The combined efficacy trials were regarded as one study at four locations. The first lactation information from the 1, 3, 5X study was allowed to be submitted separately, in order to draw preliminary conclusions before the second lactation data were submitted. The regulators examined clinical mastitis, subclinical mastitis, and somatic cell count (SCC), which measures the leukocytes (white blood cells) in milk. Somatic cell count is measured as an indication of risk for mastitis; a higher count indicates a greater risk. However, the relationship between SCC and mastitis is not necessarily direct; a high count may indicate that the immune system is protecting itself against mastitic organisms (Burton et al. 1994, 183).

16 Clinical and subclinical mastitis data from the Utah trials were excluded from the pooled analysis, since the infection was rarely treated at this site, thus confounding the effect of rbGH. The long-term animal toxicity study was analyzed separately because of the higher dose administered. Mastitis incidence was analyzed separately according to parity, controlling for parity (for this term, see note 9, above), or by ignoring parity.

17 Kronfield is professor of agriculture and of veterinary medicine at Virginia Polytechnic Institute and State University. In 1957 he began an academic appointment at the University of California, Davis, and later was appointed to the University of Pennsylvania. He conducted a small trial

with growth hormone in which cows suffered from a condition he referred to as "subclinical ketosis," which he argued would predispose cows to a number of diseases. He has not conducted long-term trials with rbGH.

18 The FDA also consulted with outside experts on the lameness study and with the USDA on injection-site reactions.

19 As noted earlier, an FDA official indicated that a similar conversation outlining what mastitis level was "approvable" may have taken place in the United States. Since the official did not reveal what that level was, I cannot determine whether it was higher – or lower – than Health Canada's cut-off point.

20 In coming to these conclusions, Monsanto had analyzed data from fifteen full-lactation trials in the United States and Europe and seventy short-term studies. Data from the Utah site, which had been excluded from the FDA evaluation, were included. The FDA, on the other hand, had analyzed eight U.S. studies, excluding Utah because of the failure to treat mastitis properly in that trial.

21 Bauman defined "catastrophic" health effects as ketosis, fatty liver, chronic wasting, crippling lameness, milk fever, mastitis, infertility, heat intolerance, sickness, suffering, and death. Mastitis was classified, therefore, as both a "catastrophic" and a "subtle" effect (1992, 3441).

22 I am grateful to Thomas MacMillan for pointing out the existence of this directive.

23 The Joint Food and Agriculture Organisation/World Health Organisation Codex Alimentarius Commission.

24 First, the EU was concerned about recent indications that the milk protein casein could protect IGF-I from digestion and that it might consequently affect the levels of growth factor in the gut and its absorption into the blood stream. Although JECFA acknowledged the protective effect of casein, it pointed out that IGF-I was still almost completely degraded within two hours. Even assuming higher IGF-I concentrations in milk, JECFA argued, absorbed IGF-I would still amount to less than 1 percent of the level circulating in the bloodstream and in the gut. Second, there were concerns that drug treatment would lead to an increased expression of retroviruses, such as bovine leukaemia virus (BLV). In response, JECFA noted that a goat experiment had not shown increased infection of cells and that BLV was destroyed by pasteurization. Third, because there were some data indicating that IGF-I stimulated prion proteins, the infective agents of BSE (mad cow disease), the EU had wanted further consideration of the possibility that rbGH treatment could shorten the incubation period for BSE. There were no data addressing this question directly; however, the committee thought a link between BSE and rbGH treatment was "highly speculative" (JECFA 1998, 9). Fourth, because exposure to

cow's milk increases a newborn's risk of developing diabetes by 1.5, the committee had been asked to consider whether treatment would lead to any additional risk, but again, JECFA argued that this was unlikely because milk from treated cows was not substantially different from that of untreated animals.

25 As noted in chapter 4, although Health Canada eventually rejected the drug on animal health grounds, the department's human health evaluation had concluded that milk from treated cows would be safe for human consumption; therefore, Canada did not oppose the adoption of a Codex standard.

CHAPTER SIX

1 A recent report has argued that rbGH is not scale-neutral. Barham, Jackson-Smith, and Mung (2000) report that "The association of rBST with other productivity-enhancing technology use helps to explain the size bias in rBST adoption. Adopters of rBST appear to have a certain production system orientation that gives rise to the use of a whole package of technologies, facilities, and management practices, most of which reward rBST use. This view is reinforced by the fact that rBST adopters in Wisconsin are also far more likely than non-adopters to be rapidly expanding their herds and investing in large-scale parlor-freestall milking operations. Since many of these high-productivity technologies, facilities, or management practices have strong technical, investment, or labor scale biases, their differential adoption profiles and their association with rBST use affect the scale neutrality of rBST adoption" (184).

Bibliography

Aaronson, Stuart A. 1991. "Growth Factors and Cancer." *Science* 254 (22 November): 1146–51.

Abbate, Gay. 1998. "Food Banks Not Filling Hunger Gap: Study." *Globe and Mail,* 13 April, A6.

Abraham, Carolyn. 1998. "Cancer-Cure Hysteria Sweeps Italy." *Globe and Mail*, 28 March, A1, A8.

Act to Amend Stevenson-Wydler Technology Innovation Act of 1980, PL 99–502.

ADA (American Dietetic Association). 1993. "The American Dietetic Association Supports Food and Drug Administration's Approval of BST." Chicago, 5 November.

– 1997. "Overview of Position Development Process." Chicago.

Allen, Susan. 1994. BGH Label Bill Passes in Senate. *Times Argus*, 19 March, 1.

American Medical Association (AMA). 1993. "AMA Supports FDA Approval of Bovine Somatotropin (BST)." News release. Chicago, 5 November.

Animal Drug Availability Act. 1996. Public Law 104-250 [H.R. 2508], 104th Congress, 2nd Session, 9 October.

American Society of Hospital Pharmacists. 1997. "AHFS Drug Information." Bethesda, MD: Board of Directors of the AHFS.

Asimov, G.J,. and N.K. Krouse. 1937. "The Lactogenic Preparations from the Anterior Pituitary and the Increase of Milk Yield in Cows." *Journal of Dairy Science* 20: 289.

"Aventis Divests Ag-Biotech Business." 2000. *Dow Jones Business News*, 15 November, at http://www.gene.ch/genet/2000/Nov/mgs00044.html.

Bagli, Charles V. 1991. "Merck Unit to Novartis for $910 Million." *New York Times*, 14 May, C7.

Barboza, David. 1999. "Is the Sun Setting on Farmers?" *New York Times*, 28 November, sections 1, 3, 14.
– 2001. "The Power of Roundup." *The New York Times*, 2 August, C1, C6.
Barham, Bradford L., Douglas Jackson-Smith, and Sunung Moon. 2000. "The Adoption of rbST on Wisconsin Dairy Farms." *AgBio Forum* 3 (nos. 2, 3): 181–7.
Barnett, Alicia Ault. 1995. "Biotech's Winning Ways on Capitol Hill." *The Lancet* 346 (14 October): 1027.
Bauman, Dale E. 1992. "Bovine Somatotropin: Review of an Emerging Animal Technology." *Journal of Dairy Science* 75: 3432–51.
Bauman, Dale E., and W. Bruce Currie. 1980. "Partitioning of Nutrients during Pregnancy and Lactation: A Review of Mechanisms Involving Homeostasis and Homeorhesis." *Journal of Dairy Science* 63: 1514–29.
Bauman, Dale E., Douglas L. Hard, Brian A. Crooker, Mary S. Partridge, Karen Garrick, Leslie D. Sandles, Hollis N. Erb, S.E. Franson, Gary F. Hartnell, and R.L. Hintz. 1989. "Long-Term Evaluation of a Prolonged-Release Formulation of N-Methionyl Bovine Somatotropin in Lactating Dairy Cows." *Journal of Dairy Science* 72: 642–51.
Bauman, Dale E., Philip J. Eppard, Melvin J. DeGeter, and Gregory M. Lanza. 1985. "Responses of High-Producing Dairy Cows to Long-Term Treatment with Pituitary Somatotropin and Recombinant Somatotropin." *Journal of Dairy Science* 68: 1352–62.
Bayh-Dole Act. 1980. Public Law 96-517, 96th Congress, 2nd session (12 December).
Baxter, Robert C. 1988. "The Insulin-Like Growth Factors and their Binding Proteins." *Comparative Biochemistry and Physiology* 91B: 229–35.
Beck, Ulrich. 1992. *Risk Society: Towards a New Modernity*. Translated by Mark Ritter. London: Thousand Oaks; New Delhi: Sage.
– 1994. "The Reinvention of Politics: Towards a Theory of Reflexive Modernization." In *Reflexive Modernization: Politics, Tradition and Aesthetics in the Modern Social Order*. Cambridge: Polity Press, in association with Blackwell Publishers.
Bifani, Paolo.1993. "The International Stakes of Biotechnology and the Patent War: Considerations After the Uruguay Round." *Agriculture and Human Values* 10 (spring): 48–53.
Bishop, J. Russell. 1993. "Statement before the Food and Drug Administration Veterinary Medicine Advisory Committee (VMAC)." *Summary Minutes of VMAC Meeting*. Gaitersburg, MD. 31 March.
Block, Elliot. 1994. "Statement before the House of Commons Standing Committee on Agriculture and Agri-Food." *Minutes of Proceedings and Evidence of the Standing Committee on Agriculture and Agri-Food Respecting Consideration of the Second Report of the Steering Committee, Pursuant to Standing Order 108(2), Consideration of Issues Relating to the Bov-*

ine Somatotropin Hormone (BST). 35th Parliament, 1st Session, 10 March.

Blumenthal, David, Michael Gluck, Karen Seashore Louis, and David Wise. 1986. "Industrial Support of University Research in Biotechnology." *Science* 231 (February): 242–6.

Bohrer, Robert A. 1994. "Food Products Affected by Biotechnology." *University of Pittsburgh Law Review* 55 (spring): 653–79.

Bradshaw, Ralph A., Ruth A. Hogue-Angeletti, and William A. Frazier. 1974. "Nerve Growth Factor and Insulin: Evidence of Similarities in Structure, Function, and Mechanism of Action." In *Recent Progress in Hormone Research: Proceedings of the 1973 Laurentian Hormone Conference*, edited by Roy O. Greep. New York and London: Academic Press.

Brammer, Rhonda. 1995. "The Right Formula." *Barron's*. 22 May, 29–36.

Brickman, Ronald, Sheila Jasanoff, and Thomas Ilgen. 1985. *Controlling Chemicals: The Politics of Regulation in Europe and the United States*. Ithaca and London: Cornell University Press.

Brinkman, George L. 1995. "U.S. Consumer Reaction to the Introduction of rBST in Milk (Consumption Patterns through January 1995)." Executive summary, prepared for the rBST Task Force, in *rBST Task Force Review of the Potential Impact of Recombinant Bovine Somatotropin (rBST) in Canada: Full Report*. Presented to the Minister of Agriculture and Agri-Food Canada, May 1995.

Brown, Paul, and John Vidal. 1999. "GM Investors Told to Sell Their Shares." *Guardian*, 25 August.

Bueckert, Dennis. 1994. "Monsanto Canada Demands Retraction from Fifth Estate." *Winnipeg Free Press*, 6 December.

– 1998. "Labelling Genetically Altered Foods Creates Split." *Globe and Mail* 30 May, A8.

Burroughs, Richard. 1993. "Statement to the Food and Drug Administration, Veterinary Medicine Advisory Committee (VMAC)." *Summary Minutes of VMAC Meeting*. Gaitersburg, MD. 31 March.

– 1994. "Statement before the House of Commons Standing Committee on Agriculture and Agri-Food." *Minutes of Proceedings and Evidence of the Standing Committee on Agriculture and Agri-Food Respecting Consideration of the Second Report of the Steering Committee, Pursuant to Standing Order 108(2), Consideration of Issues Relating to the Bovine Somatotropin Hormone (BST).* 35th Parliament, 1st Session, No. 3. 9 March.

Burton, Jeanne L., Brian W. McBride, Elliot Block, David R. Glimm, and John J. Kennelly. 1994. "A Review of Bovine Growth Hormone." *Canadian Journal of Animal Science* 74 (June): 167–201.

Butler, Declan. 1996. "Call for Reform of Scientific Panels." *Nature* 384, 12 December, 503.

"Business Notes: The Mouse That Roared." 1998. *MacLean's Magazine*, 4 May, 42.

Campbell, Murray. 1998a. "Fears about Food Safety on the Rise." *Globe and Mail* 13 April, A1, A7.

– 1998b. "Food-Borne Toxins Take on Dangerous New Forms." *Globe and Mail* 14 April, A1, A4.

Canada. Agriculture and Agri-Food Canada. Policy Branch. 1995. "Economic Impacts of rbST in Canada: Prepared for the rbST Task Force." In *rbST Task Force Review of the Potential Impact of Recombinant Bovine Somatotropin (rbST) in Canada: Full Report*. Presented to the Minister of Agriculture and Agri-Food Canada, May.

Canada. CDC. (Canadian Dairy Commission). 1997. News release.

– CHC. (Canadian Health Coalition). 1998. *A Citizens' Guide to the Health Protection Branch Consultation*, at http://www.healthcoalition.ca/ abdication.html

– Department of Finance. *1996 Budget Plan*. Ottawa, 6 March.

– *Food and Drugs Act*. 1985. R.S. 1985, C.F-27.

Canada Gazette. 1995. Part 2, vol. 129, no. 14 (12 July): 1779.

Canada. Government of Canada. 1994. *Government Response to the Report of the Standing Committee on Agriculture and Agri-Food, rbST in Canada*. Ottawa, August.

– 1998a. *Canadian Biotechnology Strategy: Summary of Round Table Conclusions*. July.

– 1998b. "Federal Government Releases New Biotechnology Strategy." News release, 6 August.

Canada. Health Canada. 1995a. "Statement on Bovine Somatotropin (rbST), Prepared for the rbST Task Force." In *rbST Task Force Review of the Potential Impact of Recombinant Bovine Somatotropin (rbST)* In *Canada: Full Report*. Presented to the Minister of Agriculture and Agri-Food Canada, May, appendix 2–3.

– 1995b. *Response to the Motion of the Standing Committee on Agriculture and Agri-Food regarding rbST*, 21 June1.

– 1998a. *Health Protection for the Twenty-first Century: A Discussion Paper*, July, at http://www.hc-sc.gc.ca/hpb/transitn/index.html.

– 1998b. *Health Protection Branch Transition: Legislative Renewal*, at http://www.hc-sc.gc.ca/hpb/transitn/3pager_e.html.

– 1998c. *Health Protection Branch Transition: Risk Management*, at http://www.hc-sc.gc.ca/hpb/transitn/rmf-e.html.

– 1999a. "Health Canada Rejects Bovine Growth Hormone in Canada." News release. 14 January, at http://www.hc-sc.gc.ca/english/archives/releases/99-03e.htm.

– 1999b. *Health Protection in the New Food Safety and Inspection Bill*, at http://www.hc-sc.gc.ca/food-aliment/english/subjects/food-bill.html.

Canada. House of Commons. Standing Committee on Agriculture and Agri-Food. 1994a. *Minutes of Proceedings and Evidence of the Standing Committee on Agriculture and Agri-Food Respecting Consideration of the Second Report of the Steering Committee, Pursuant to Standing Order 108(2), Consideration of Issues Relating to the Bovine Somatotropin Hormone (bST).* 35th Parliament, 1st Session, 3–8 November, 7–10, 15 March.

– 1994b. *rbST in Canada.* Ottawa, April.

Canada. House of Commons. Standing Committee on Environment and Sustainable Development. 1995a. *Minutes of Proceedings and Evidence of the Standing Committee on Environment and Sustainable Development, Review of the Canadian Environmental Protection Act.* Nos. 34–78, 14 June 1994–March 1995.

– 1995b. *It's about Our Health! Towards Pollution Prevention:* CEPA *Revisited. Report of the House of Commons Standing Committee on Environment and Sustainable Development Pursuant to its Order of Reference of 10 June 1994.*

Canada. Industry Canada. 1993. "Federal Government Agrees on New Regulatory Framework for Biotechnology." News release, Ottawa, 11 January.

– 1994. *Building a More Innovative Economy.* Ottawa, November.

– 1998. "Minister Manley Announces Proposed Changes to Canada's Drug Patent Regulations." News release, 21 January.

Canada. Minister of Finance. 1996. "Budget Plan: Including Supplementary Information and Notices of Ways and Means Motions." 6 March.

– NBAC (National Biotechnology Advisory Committee). 1998. *Sixth Report – Leading in the Next Millennium.* Ottawa.

– Senate of Canada. 1998a. *Minutes of Evidence Presented before the Standing Senate Committee on Agriculture and Forestry (Unrevised).* Ottawa, 4 June, 22 October, 29 October, 17 November, 7 December.

– 1999a. *rbST and the Drug Approval Process.* Interim Report of the Standing Senate Committee on Agriculture and Forestry, March.

– 1999b. *Minutes of Evidence Presented Before the Standing Senate Committee on Agriculture and Forestry (Unrevised).* Ottawa, 26 April, 3 May, 13 May.

Canadian Environmental Protection Act (CEPA). 1985. R.S. 1985 c. 15 (4th Supp.).

Canadian Healthcare Association. 1998. "Submission to the Health Protection Branch Public Consultation on HPB Transition," October, at http://www.canadian-healthcare.org/chapol/transition.htm.

Canadian Health Coalition. 1998. "Transition = Abdication: A Citizens' Guide to the Health Protection Branch Consultation (HPB Transition)," September, at http://www.healthcoalition.ca/abdication.html.

CFIA (Canadian Food Inspection Agency). 1998. *Canada and United States Bilateral on Biotechnology*, at http://inspection.gc.ca/english/plaveg/pbo/usda01e.shtml.

Caplan, Barbara G. 1994. "Letter to Minister of Health Diane Marleau," 1 June. Toronto Food Policy Council Archives.

Carey, John. 1998. "We Are Now Starting the Century of Biology." *Business Week*, 31 August, 86–7.

CAST (Council for Agricultural Science and Technology). 1993. News release, Ames, IA, 27 May.

Cattell, Meg. 1996. "Statement to the Food and Drug Administration Veterinary Medicine Advisory Committee." *Transcript of Proceedings to Review 28 Herd Post-Approval Monitoring Study and Findings from the Commercial Use of Sometribove*. Rockville, MD, 20 November.

Centner, Terence J., and Kyle W. Lathrop. 1997. "Labeling rBST-Derived Milk Products: State Responses to Federal Law." *University of Kansas Law Review* 45: 511–56.

Challacombe, D.N., and E.E. Wheeler. 1994. "Safety of Milk from Cows Treated with Bovine Somatotropin" (letter). *The Lancet* 344 (17 September): 815–16.

Chan, June M., Meir J. Stampfer, Edward Giovannucci, Peter H. Gann, Jing Ma, Peter Wilkinson, Charles H. Hennekens, and Michael Pollak. 1998. "Plasma Insulin-Like Growth Factor-I and Prostate Cancer Risk: A Prospective Study." *Science* 279 (23 January): 563–6.

Chase, Stacey. 1996. "Court Reverses BST Law." *Burlington Free Press*, 9 August.

Christiansen, Andrew. 1995. *Recombinant Bovine Growth Hormone: Alarming Tests, Unfounded Approval*. Montpelier, VT: Rural Vermont.

CIELAP (Canadian Institute of Environmental Law and Policy). 1994. "A Broad Coalition of Public Interest Groups Urges the Prime Minister to Delay Growth Hormone for Cows." News release, 10 August.

CIHR (Canadian Institutes of Health Research). 2000. "Who We Are," at http://www.cihr.ca/welcome_e.shtml

City of Toronto. Board of Health. 1994a. "Report No. 04, Clause 5, The Current Status of the Licensing of Recombinant Bovine Growth Hormone (rbGH)," submitted 21 February. Toronto Food Policy Council Archives.

– 1994b. "Deputation List." 28 April. Toronto Food Policy Council Archives.

– 1994c. "Memo to Toronto Food Policy Council from Marbeth Greer." 2 May. Toronto Food Policy Council Archives.

Clinton, Bill, and Al Gore. 1996. "Reinventing the Regulation of Animal Drugs: National Performance Review." Washington, DC, May. Photocopy.

Coalition of the Americas. 1997. "Food Standard Issues to Be Considered by the Codex Alimentarius." Brochure presented in coordination with the Confédération Mondiale De L'Industrie De La Santé Animal (COMISA).

Codex Alimentarius Commission. No date. "General Decisions of the Codex Alimentarius Commission: Statements of Principle Concerning the Role of Science in the Codex Decision-Making Process and the Extent to

Which Other Factors Are Taken into Account, at http://www.codexalimentarius.net/manual/decide.htm.
– 1997a. "Matters Relating to the Implementation of the WTO Agreements on the Application of Sanitary and Phytosanitary Measures and the Agreement on Technical Barriers to Trade." Twenty-Second Session, International Conference Centre, Geneva, 23–28 June, Agenda Item 11, 21 May.
– 1997b. Twenty-Second Report, Draft Maximum Residue Limits for Bovine Somatotropin (BST), Alinorm 95/31, Appendix 2, Comments from Consumers International Alinorm 97/25, Part 10, and from the European Community CAC/LIM 17.
– 1997c. Twenty-Second Session, 23–28 June 1997, Agenda Item 7, Geneva, Codex Committee on Residues of Veterinary Drugs in Food, Comments at Step 8 by the European Community on Bovine Somatotropin, Alinorm 97/25, Part 10, Addendum 2.
– 1999a. Understanding the Codex Alimentarius Commission: Preface, at http://www.fao.org/docrep/w9114e/w9114e00.htm.
– 1999b. Joint WHO/WHO Food Standards Programme, Codex Alimentarius Commission Twenty-Third Session, Rome, 28 June–3 July, Agenda Item 9, Consideration of Draft Standards and Related Texts, Maximum Residue Limits for Bovine Somatotropin, Alinorm 99/21, Part 1, Addendum 2, May 1999.
– 2001. Joint FAO/WHO Food Standards Programme, Codex Alimentarius Commission Twenty-Fourth Session, Geneva, 2–7 July, 2001, Report of the Fifteenth Session of the Codex Committee on General Principles, Paris, France, 10–14 April 2000, Alinorm 01/33 at ftp://ftp.fao.org/codex/ALINORM01/A101_33e.pdf
Coelho, Tony. 1987. "Statement to the U.S. House of Representatives Subcommittee on Livestock, Dairy, and Poultry of the Committee on Agriculture." *Review of Status and Potential Impact of Bovine Growth Hormone: Hearing before the Subcommittee on Livestock, Dairy, and Poultry of the Committee on Agriculture*. 99th Congress, 2nd Session, Serial no. 99-51, 11 June. Washington, DC: U.S. Government Printing Office.
Coghlan, Andy. 1994. "Keep Milk Hormone Ban, Say Farmers." *New Scientist*, 13 August, 8.
– 1996. "Europe Pushes to Ban Antibiotic Down on the Farm." *New Scientist*, 21–28 December, 6.
Collier, Robert J. 1993. "Statement to the Food and Drug Administration Veterinary Medicine Advisory Committee (VMAC)." *Summary Minutes of VMAC Meeting*. Gaitersburg, MD. 31 March.
– 1994. "Statement before the House of Commons Standing Committee on Agriculture and Agri-Food." *Minutes of Proceedings and Evidence of the Standing Committee on Agriculture and Agri-Food Respecting Consideration of the Second Report of the Steering Committee, Pursuant to Standing*

Order 108(2), Consideration of Issues Relating to the Bovine Somatotropin Hormone (BST). 35th Parliament, 1st Session, No. 3. 10 March.

Collier, Robert J., David R. Clemmons, and Sharon M. Donovan. 1994. "Safety of Milk from Cows Treated with Bovine Somatotropin" (letter). *The Lancet* 344 (17 September): 816.

Condon, Robert. 1996. "Statement to the Food and Drug Administration Veterinary Medicine Advisory Committee (VMAC)." Transcript of Proceedings to Review 28 Herd Post-Approval Monitoring Study and Findings from the Commercial Use of Sometribove. Rockville, MD, 20 November. Washington, DC: Miller Reporting Company.

COMISA (Confédération Mondiale de l'Industrie de la Santé Animale). 1997. "Adopting Standards for the veterinary Drug bST: The Case for Science-Based Codex Standards." A Codex briefing document.

Consumers' Union. 1995. "Frivolous Lawsuits Sink to New Low as Monsanto Sues Small, Family-Owned Dairy." News release, 14 June.

Council of Canadians. 1995a. "A Critique of the BST Review: Selected Briefing Notes Documenting the Inaccurate and Biased Information Presented to Members of Parliament by Government Officials and BST Manufacturer Monsanto." Toronto Food Policy Council Archives.

– 1995b. "Letter to Bob Speller," 9 June. Toronto Food Policy Council Archives.

– 1995c. Media Release, 16 June. Toronto Food Policy Council Archives.

Council of the European Union. 1998. "Council Directive 98/58/EC of 20 July 1998 Concerning the Protection of Animals Kept for Farming Purposes," 20 July, Official Journal L221: 0023-0027, at http://europa.eu.int/eur-lex/en/lif/dat/1998/en_398L0058.html

CSA (Council on Scientific Affairs of the American Medical Association). "Biotechnology and the American Agricultural Industry." *Journal of the American Medical Association* 265 (20 March): 1429–36.

Culliton, B.J. 1977. "Harvard and Monsanto: The $23-million Alliance." *Science* (25 February): 759–63.

Curry, Paul. 1981. "Statement to the House of Representatives Subcommittee on Investigations and Oversight, and the Subcommittee on Science, Research, and Technology of the Committee on Science and Technology." *Commercialization of Academic Biomedical Research: Hearings Before the Subcommittee on Investigations and Oversight, and the Subcommittee on Science, Research, and Technology of the Committee on Science and Technology*. 97th Congress, 1st Session, 8–9 June.

CVMA (Canadian Veterinary Medical Association) Expert Panel on rbST. 1998. "Report of the Canadian Veterinary Medical Association Expert Panel on rbST," prepared for Health Canada, November, at http://www.hc-sc.gc.ca/english/archives/rbst/animals/index.htm

Czech, Michael P. 1989. "Signal Transmission by the Insulin-Like Growth Factors." *Cell* 59 (20 October): 235–8.

Daniel, Victor. 1996. "Recombinant Bovine Growth Hormone – The Canadian Opportunity: Why Rejecting the Hormone is in the National Interest." Unpublished paper.

Daughaday, William H., and David M. Barbano. 1990. "Bovine Somatotropin Supplementation of Dairy Cows: Is the Milk Safe? Special Communication." *Journal of the American Medical Association* 264 (22–29 August): 1003–5.

Dekkers, J.C.M., H.W. Leach, E.B. Burnside, and A.H. Fredeen. 1995. "Assessment of the Potential Impact of the Use of rbST on the Genetic Evaluation and Improvement of Dairy Cattle in Canada." Executive Summary prepared for the rbST Task Force. In *rbST Task Force Review of the Potential Impact of Recombinant Bovine Somatotropin (rbST) in Canada: Full Report*, presented to the Minister of Agriculture and Agri-Food Canada, May 1995.

Department of Health and Human Services. 1993. *News release*, Washington, DC, 5 November, 2.

Deutsch, Claudia H. 1998. "Sticking to the Formula." *The New York Times*, 3 March, C1, C3.

Ditchburn, Jennifer. 1998. "Health Canada Condemned." *Globe and Mail* 8 July, A4.

Djerassi, Carl. 1981. "Statement to the Subcommittee on Investigations and Oversight and the Subcommittee on Science, Research and Technology of the Committee on Science and Technology, U.S. House of Representatives." *Commercialization of Academic Research: Hearings before the Subcommittee on Investigations and Oversight and the Subcommittee on Science, Research and Technology of the Committee on Science and Technology*, 97th Congress, 1st Session, 8–9 June.

Doyle, Jack. 1985. *Altered Harvest: Agriculture, Genetics, and the Fate of the World's Food Supply*. New York: Viking Press.

"The Drama in Cow's Milk." 1998. *Globe and Mail*, 19 September, D6.

Drug Export Amendments Act. 1986. Public Law 99-660 [S. 17477] 99th Congress 2nd Session, 14 November.

"Drug Residues Found in Milk." 1988. *New York Times*, 14 April, A32.

Duncan, Colin. 1997. *The Centrality of Agriculture: Between Humankind and the Rest of Nature*. Montreal and Kingston: London, McGill-Queens University Press.

Economic Recovery Taxation Act. 1981. Public Law 97-34 [H.R. 4242], 97th Congress, 1st Session, 13 August.

Eggertson, Laura. 1997a. "Cow Hormone Smuggling Foiled." *Globe and Mail*, 3 February, B1, B5.

– 1997b. "Drug-Approval Process Criticized." *Globe and Mail*, 28 May, A1, A10.

– 1998a. "Health Officials Trying to Kill Critical Parts of Drug Report." *Toronto Star*, 29 May, A6.

– 1998b. "Rock Denies Cover-Up on Milk-Boosting Drug." *Toronto Star*, 30 May, A16.
– 1998c. "Researchers Threatened, Inquiry Told." *Toronto Star*, 17 September, A6.
– 1998d. "Adviser Had No Conflict, College Says." *Toronto Star*, 24 September, A8.
Eisenberg, Rebecca S. 1987. "Proprietary Rights and the Norms of Science in Biotechnology Research." *Yale Law Journal* 97 (December): 186–97.
– 1994. "A Technology Policy Perspective on the NIH Gene Patenting Controversy." *University of Pittsburgh Law Review* 55 (spring): 633–52.
Envision Research. 1997. *Socioethical Implications of Biotechnology*. Kanata, ON, sponsored by the Department of Western Economic Diversification.
Eppard, P.J., L.A. Bentle, B.N. Violand, S. Ganguli, R.L. Hintz, L.Kung Jr, G.G. Krivi, and G.M. Lanza. 1992. "Comparison of the Galactopoietic Response to Pituitary-Derived and Recombinant-Derived Variants of Bovine Growth Hormone." *Journal of Endocrinology* 132 (January): 47–56
Epstein, Samuel S. 1989. "BST: the Public Health Hazards." *The Ecologist* 19: 191–5.
– 1990a. "Potential Public Health Hazards of Biosynthetic Milk Hormones." *International Journal of Health Services* 20: 73–84.
– 1990b. "Questions and Answers on Synthetic Bovine Growth Hormones." *International Journal of Health Services* 20: 573–82.
– 1991. "A Reply to Virginia Weldon." *International Journal of Health Services* 21: 563–4.
– 1994. "A Needless New Risk of Breast Cancer (Commentary)." *Los Angeles Times*, 20 March, M5.
– 1996. "Unlabeled Milk from Cows Treated with Biosynthetic Growth Hormones: A Case of Regulatory Abdication." *International Journal of Health Services* 26: 173–85.
"Europe to Cut Butter Stocks." 1987. *New York Times*, 11 February, D2.
European Commission. 1997. *Commission Decision Setting Up Scientific Committees in the Field of Consumer Health and Food Safety*, DG 24, Commission Decision Number 97/579/EC, 23 July.
European Community. 1997. *Comments at September 8 by the European Community on Bovine Somatotropin, Codex Committee on Residues of Veterinary Drugs in Food*, Codex Alimentarius Commission Twenty-second Session, 23–28 June, Geneva.
Ewing, Terzah. 1996. "Monsanto Plans to Cut up to 2,500 Jobs After Spinoff of Its Chemical Business." *Wall Street Journal* 10 December, A4.
Expert Panel on the Commercialization of University Research. 1999. *Public Investment in University Research: Reaping the Benefits*. Report of the Expert Panel on the Commercialization of University Research, presented to the prime minister's Advisory Council on Science and Technology, 4 May, 1999.

Fagan, Drew. 1996. "Canada Triumphs in Tariff Battle." *Globe and Mail* 16 July, A1, A8.

FAO (Food and Agriculture Organization). 1997. "European Ban on Hormone-Treated Cattle Rejected by WTO." News release, http://www.fao.org.

"Farmers Finding a Voice in New Vermont Group." 1986. *New York Times*, 14 December, section 1, 72.

Farnsworth, Clyde H., and David Binder. 1988. "Briefing: Supports Questioned." *New York Times*, 23 December, 126.

Faulkner, Wendy, Jacqueline Senker, and Lea Velho. 1995. *Knowledge Frontiers: Public Sector Research and Industrial Innovation in Biotechnology, Engineering Ceramics, and Parallel Computing*. Oxford: Clarendon Press.

FDA (Food and Drug Administration). 1988. A Technical Assistance Document for Efficacy Studies of Bovine Somatotropin (BST) in Lactating Dairy Cows. Draft. 24 March, 1–13.

– 1992. "Statement of Policy: Foods Derived from New Plant Varieties (Notice)." *Federal Register* 57, no. 104 (29 May) 22984.

– 1993a. *Joint Meeting of the Food Advisory Committee and Veterinary Medicine Advisory Committee: Transcript*. Washington, DC, 6 May, Miller Reporting.

– 1993b. "Animal Drug, Feeds, and Related Products; Sterile Sometribove Zinc Suspension." *Federal Register* 58, no. 217 (12 November): 59946–7.

– 1993c. "Food Labeling: Foods Derived from New Plant Varieties." *Federal Register* 88, no. 8 (28 April): 25837.

– 1993d. *Freedom of Information Summary,* NADA *140-872 Posilac (Sterile Sometribove Zinc Suspension),* at http://www.cvm.fda.gov/fda/TOCs/bsttoc.html.

– 1994a. "Interim Guidance on the Voluntary Labeling of Milk and Milk Products from Cows That Have Not Been Treated with Recombinant Bovine Somatotropin." *Federal Register* 59, no. 28 (10 February): 6279–80.

– 1994b. "Talk Paper: Dietary IGF-1 and rbST". T94-17.

– 1999. "Report on the Food and Drug Administration's Review of the Safety of Recombinant Bovine Somatotropin," at 10 February, at http://www.fda.gov/cvm/index/bst/RBRPTFNL.htm.

Feder, Barnaby J. 1993. "In Battling for Biotech, Monsanto Is the Leader." *New York Times* 24 December, D1, D4.

– 1996. "Out of the Lab: A Revolution on the Farm." *New York Times* 3 March, F1, F11.

Federal Court of Canada. 2000. *Haydon v. Canada*, at http://www.fja.gc.ca/en/cf/2000/orig/html/2000fca27277.o.e.html.

Federal Technology Transfer Act. 1986. Public Law 99-502, 99th Congress (20 October).

Flynn, Julia, John Carey, and William Echikson. 1998. "The Pros and Cons of Gene-Spliced Food." *Business Week*, 2 February, 62–3.

Follet, Ghislain. 2000. "Antibiotic Resistance in the EU – Science, Politics and Policy." *AgBioForum*, 3, nos. 2, 3 (spring/summer): 148–55.

Food and Drug Law Group, Blake, Cassels and Graydon. 1994. "Developments in Canadian Law Relating to Food, Drugs, Devices and Cosmetics as of December 1992." *Food and Drug Law Journal* 49: 323–57.

Food Chemical News. 1992. (10 February) 76.

Foster, Kent. 1994. "Statement before the House of Commons Standing Committee on Agriculture and Agri-Food." *Minutes of Proceedings and Evidence of the Standing Committee on Agriculture and Agri-Food Respecting Consideration of the Second Report of the Steering Committee, Pursuant to Standing Order 108(2), Consideration of Issues Relating to the Bovine Somatotropin Hormone (BST)*. 35th Parliament, 1st Session, No. 3. 7 March.

Foster, Scott. 2000. "Supreme Court to Enter Debate over Biotech Mouse," *Capital News Online*, at http://temagami.carleton.ca/jmc/cnews/20102000/n3.htm.

Freeman, Christopher, and Carlota Perez. 1988. "Structural Crises of Adjustment, Business Cycles, and Investment Behaviour." In *Technical Change and Economic Theory*, edited by Giovanni Dosi, Christopher Freeman, Richard Nelson, Gerald Silverberg, and Luc Soete. London and New York: Pinter Publishers.

Friedmann, Harriet. 1991. "Changes in the International Division of Labor: Agri-food Complexes and Export Agriculture." In *Towards a New Political Economy of Agriculture* edited by William H. Friedland, Lawrence Busch, Frederick H. Buttel, and Alan P. Rudy. Oxford: Westview Press.

– 1994. "Distance and Durability: Shaky Foundations of a World Food Economy." In *The Global Restructuring of Agro-Food Systems*, edited by Philip McMichael. Ithaca and London: Cornell University Press.

GAO (General Accounting Office). 1990. *Food Safety and Quality: FDA Surveys Not Adequate to Demonstrate Safety of Milk Supply*. RCED-91-26. Washington, DC: Government Printing Office.

– 1992a. *Food Safety and Quality: FDA Needs Stronger Controls Over the Approval Process for New Animal Drugs*. RCED-92-63. Washington, DC: Government Printing Office.

– 1992b. *Food Safety and Quality: FDA Strategy Needed to Address Animal Drug Residues in Milk*. RCED-92-209. Washington, DC: Government Printing Office.

– 1992c. *Recombinant Bovine Growth Hormone: FDA Approval Should Be Withheld Until the Mastitis Issue is Resolved*. PEMD-92-26. Washington, DC: Government Printing Office.

– 1994. Enclosure to Letter to George E. Brown Jr. Summary of Results B-257122. Rural Vermont Archives.

Gershon, Diane. "Monsanto Sues over BST." *Nature* 368 (31 March): 384.

Geyer, Richard E. 1997. "Extralabel Drug Use and Compounding in Veterinary Medicine." *Food and Drug Law Journal* 52: 291–5.

Gibbons, Ann. 1990. "FDA Publishes Bovine Growth Hormone Data." *Science* 249: (24 August): 852–3.

Goldberg, Jeffrey. 1996. "Next Target: Nicotine." *New York Times Magazine*, 4 August, 22–36.

Goodman, David. 1991. "Some Recent Tendencies in the Industrial Reorganization of the Agri-food System." In *Towards a New Political Economy of Agriculture* edited by William H. Friedland, Lawrence Busch, Frederick H. Buttel, and Alan P. Rudy. Oxford: Westview Press.

– 1997. "World-Scale Processes and Agro-Food Systems: Critique and Research Needs." *Review of International Political Economy* 4, no. 4 (winter) 663–87.

Goodman, David, Bernardo Sorj, and John Wilkinson. 1987. *From Farming to Biotechnology: A Theory of Agro-Industrial Development*. Oxford and New York: Basil Blackwell.

Gougeon, Rejeanne, and Arlene Taveroff. 1994. "Recombinant Bovine Somatotropin Supplementation of Cows and Wholesomeness of Milk." *Journal of the Canadian Dietetic Association* 55 (fall): 143–5.

Gravelle, Louis-Pierre, and Zhen Wong. 2000. "Federal Court of Appeals Rules OncoMouse Patentable," at http://www.robic.ca/publications/142-117.htm.

Gray, Paul.1981. "Statement to the U.S. House of Representatives Subcommittee on Investigations and Oversight and the Subcommittee on Science, Research and Technology of the Committee on Science and Technology, U.S. House of Representatives." *Commercialization of Academic Research: Hearings before the Subcommittee on Investigations and Oversight and the Subcommittee on Science, Research and Technology of the Committee on Science and Technology, U.S. House of Representatives*. 97th Congress, 1st session, 8–9 June.

Groenewegen, Paul P., Brian W. McBride, John H. Burton and Theodore H. Elsasser. 1990. "Bioactivity of Milk from bST-Treated Cows." *Journal of Nutrition* 120 (May): 514–20.

Guyda, Harvey. 1994. "Statement before the House of Commons Standing Committee on Agriculture and Agri-Food." *Minutes of Proceedings and Evidence of the Standing Committee on Agriculture and Agri-Food Respecting Consideration of the Second Report of the Steering Committee, Pursuant to Standing Order 108(2), Consideration of Issues Relating to the Bovine Somatotropin Hormone (BST)*. 35th Parliament, 1st session, no. 3. 8 March.

Ha, Tu Thanh. 1994. "Conflict Questions Raised Over Official's Milk Testimony." *Globe and Mail*, 7 December, A5.

Halpern, Sue M. 1988. "At 35 cents an Hour, How're Ya Gonna Keep 'em Down on the Dairy Farm?" *New York Times* 11 June, section 1, 34.

Hammond, B.G., R.J. Collier, M.A. Miller, M. McGrath, D.L. Hartzell, C. Kotts, and W. Vandaele. 1990. "Food Safety and Pharmacokinetic Studies

Which Support a Zero (0) Meat and Milk Withdrawal Time for Use of Sometribove in Dairy Cows." *Annales Recherches Veterinaires*, supp. 1, 107s–120s.

Hansen, Michael K. 1990. *Biotechnology and Milk: Benefit or Threat? An Analysis of Issues Related to bGH/bST Use in the Dairy Industry*. Mt. Vernon, New York: Consumer Policy Institute, Consumers' Union.

– 1993. *Testimony before the Veterinary Medicine Advisory Committee on Potential Animal and Human Health Effects of rbGH Use*. New York: Consumer Policy Institute, Consumers' Union, 31 March.

– 1994. "Statement before the House of Commons Standing Committee on Agriculture and Agri-Food." *Minutes of Proceedings and Evidence of the Standing Committee on Agriculture and Agri-Food Respecting Consideration of the Second Report of the Steering Committee, Pursuant to Standing Order 108(2), Consideration of Issues Relating to the Bovine Somatotropin Hormone (BST)*. 35th Parliament, 1st session, no. 3. 10 March.

Hansen, Michael. 1999. "Consumer International Press Release: Countries Free to Set Own Standards on BST Residues in Food, June 30," at http://www.gene.ch/genet/1999/Jul/msg00008.html.

Hanson, David. 1996. "Regulation after Delaney." *Chemical and Engineering News*, 23 September, 38–9.

Harper, Tim. 1997. "Cuts Pushing Top Scientists to Flee Canada." *Toronto Star* 5 October, A1, A5.

Harsanyi, Zolt. 1981. "Statement before the U.S. House of Representatives Subcommittee on Investigations and Oversight, and the Subcommittee on Science, Research, and Technology of the Committee on Science and Technology." *Commercialization of Academic Biomedical Research: Hearings before the Subcommittee on Investigations and Oversight, and the Subcommittee on Science, Research, and Technology of the Committee on Science and Technology*. 97th Congress, 1st session, 8–9 June.

Hawkey, Peter M. 1998. "Action against Antibiotic Resistance: No Time To Lose." *The Lancet* 351 (2 May): 1298–9.

Health Canada. 2000. *Budget 2000 Information: Protecting the Health of Canadians,* 28 February, at http://www.hc-sc.gc.ca/budget/english/2000/protection.htm.

"Health Monitor." 1998a. *Maclean's Magazine*, 2 February, 58.

Hightower, Jim. 1973. *Hard Tomatoes, Hard Times: A Report of the Agribusiness Accountability Project on the Failure of America's Land Grant College Complex*. Cambridge, MA: Schenkman Publishing.

Hilts, Philip J. 1990a. "U.S. Calls Milk Free of Antibiotics." *New York Times* 6 February, C13.

– 1990b. "FDA Chemist Asserts Agency Is Stalling on Tests for Milk Purity." *New York Times*, 7 February, A22.

– 1990c. "FDA Contradicts a Report on Milk." *New York Times*, 18 April, A21.

– 1995. “Ruling Gives Drug Makers up to 3 Extra Years on Patents.” *New York Times*, 26 May, C1, C5.

Hiss, Tony. 1994. “How Now, Drugged Cow? Biotechnology comes to Rural Vermont.” *Harper's Magazine* 289 (October): 80–90.

Hung, Huynh, and Michael Pollak. 1993. “Synergistic Reduction of Expression of the IGF-I Gene by the Combination of Somatostatin and Tamoxifen.” *Proceedings of ASCO* 12 (March): 121.

– 1994. “Uterotrophic Actions of Estradiol and Tamoxifen Are Associated with Inhibition of Uterine Insulin-Like Growth Factor Binding Protein 3 Gene Expression.” *Cancer Research* 54 (15 June): 3115–9.

Image Base. 1993. *BST & Milk: Issues and Answers*. Video produced in cooperation with American Medical Television.

“In Defence of the FDA.” 1995. *The Lancet* 346 (14 October): 981.

Intellectual Property Policy Directorate. 1998. *Background Economic Study of the Canadian Biotechnology Industry*, at http://strategis.gc.ca.

Industry Canada (Bio-Industries Branch). 1998. Building the Canadian Biotechnology Strategy, at http://strategis.ic.gc.ca/SSG/bh00237e.html

Jasanoff, Sheila. 1986. *Risk Management and Political Culture: A Comparative Study of Science in the Policy Context*. New York: Russell Sage Foundation.

– 1990. *The Fifth Branch: Science Advisors as Policy Makers*. Cambridge, MA, and London: Harvard University Press.

– 1995a. “Product, Process, or Programme: Three Cultures and the Regulation of Biotechnology.” In *Resistance to Technology: Nuclear Power, Information Technology, and Biotechnology*, edited by Martin Bauer. Cambridge: University of Cambridge Press.

– 1995b. *Science at the Bar: Law, Science, and Technology in America*. Cambridge, MA, and London: A Twentieth Century Fund Book, Harvard University Press.

JECFA. Joint FAO/WHO Expert Committee on Food Additives. 1999. “Residues of Some veterinary Drugs in Animals and Foods.” Monograph prepared by the fiftieth meeting of JECFA, Rome, 17–26 February, 1998.

Johnson, George. 1997. “Designing Life: Proteins 1, Computer 0.” *New York Times* 25 March, B9, B11.

Johnsrude, Larry. 1995. “Keep Milk BST-Free.” *Winnipeg Free Press* 6 April.

Joint FAO/WHO Expert Committee on Food Additives (JECFA). 1993. “Fortieth Report: Evaluation of Certain Veterinary Drug Residues in Food.” *WHO Technical Report Services* 832: 41.

Jorgensen, Neal. 1987. “Statement to the U.S. House of Representatives Subcommittee on Livestock, Dairy, and Poultry of the Committee on Agriculture.” *Status and Review of Bovine Growth Hormone: Hearing before the Subcommittee on Livestock, Dairy, and Poultry of the Committee on Agri-*

culture. 99th Congress, Second Session, 11 June, Washington DC: U.S. Government Printing Office.

Junne, Gerd. 1993. "Agricultural Biotechnology: Slow Applications by Large Corporations." *Agriculture and Human Values* 10 (spring): 40–6.

Juskevich, Judith C., and C. Greg Guyer. 1990. "Bovine Growth Hormone: Human Food Safety Evaluation." *Science* 249 (24 August): 875–84.

– 1991. "Safety of Bovine Growth Hormone (letter in reply to David S. Kronfeld)." *Science* 251 (18 January): 256.

Kalter, Robert, Robert Milligan, William Lesser, William Magrath, Loren Tauer, and Dale Bauman. 1985. *Biotechnology and the Dairy Industry: Production Costs, Commercial Potential, and the Economic Impact of the Bovine Growth Hormone*. Report Prepared for Cornell University Center for Biotechnology by the Department of Agricultural Economics and Department of Animal Science, Cornell University, Ithaca, New York, December.

– 1987. Statement to the U.S. House of Representatives Subcommittee on Livestock, Dairy, and Poultry of the Committee on Agriculture. *Status and Review of Bovine Growth Hormone: Hearing before the Subcommittee on Livestock, Dairy, and Poultry of the Committee on Agriculture*. 99th Congress, Second Session, 11 June, Washington DC: U.S. Government Printing Office.

Kastel, Mark. 1995. *Down on the Farm: The Real BGH Story*. Montpelier, Vermont: Rural Vermont.

Kay, Lily E. 1995. "A Book of Life? How a Genetic Code Became a Language." Paper presented at the Notre Dame Genome Conference, October, 1995. Also in *Controlling our Destinies: Historical, Philosophical, Social and Ethical Perspectives on the Human Genome Project*, edited by Phillip Sloan, Notre Dame, IN: University of Notre Dame Press, forthcoming 1999.

Kennedy, Donald. 1981. "Statement to the Subcommittee on Investigations and Oversight and the Subcommittee on Science, Research and Technology of the Committee on Science and Technology, U.S. House of Representatives." *Commercialization of Academic Research: Hearings before the Subcommittee on Investigations and Oversight and the Subcommittee on Science, Research and Technology of the Committee on Science and Technology*. 97th Congress, 1st Session, 8–9 June.

Kenney, Martin. 1986. *Biotechnology: The University-Industrial Complex*. New Haven, CT, and London: Yale Univeristy Press.

Kenney, Martin, Linda M. Lobao, James Curry, and W. Richard Goe. 1991. "Agriculture in U.S. Fordism: The Integration of the Productive Consumer." In *Towards a New Political Economy of Agriculture*, edited by William H. Friedland, Lawrence Busch, Frederick H. Buttel, and Alan P. Rudy. Oxford: Westview Press.

Kessler, David A., Michael R. Taylor, James H. Maryanski, Eric L. Flamm, and

Linda S. Kahl. 1992. "The Safety of Foods Developed by Biotechnology." *Science* 256 (26 June): 1747–1832.

Kilman, Scott. 1998. "Merger Changes the Crop Biotech Field." *Wall Street Journal* 3 June, B10.

Kilman, Scott. 1999a. "Monsanto is Sued over Genetically Altered Crops," *Wall Street Journal*, 15 December, A3, A10.

– 1999b. "Antitrust Regulators Are Investigating Monsanto's Control of Cotton Genes." *Wall Street Journal*, 17 December, B12

Kilman, Scott, and Thomas M. Burton. 1999. "Farm and Pharma: Monsanto Boss's Vision of 'Life Sciences' Firm Now Confronts Reality." *Wall Street Journal*, December 21, A1 & A10

Kilman, Scott, and Susan Warren. 1998. "Cargill, Monsanto in Biotech Venture." *Globe and Mail* 15 May, B8.

Kilpatrick, Andrew L. 1995. "Pure Milk, Monsanto Settle Advertising Suit." *Waco Tribune-Herald 16* June.

King, Jonathan. 1981. "Statement to the Subcommittee on Investigations and Oversight and the Subcommittee on Science, Research and Technology of the Committee on Science and Technology, U.S. House of Representatives." *Commercialization of Academic Research: Hearings before the Subcommittee on Investigations and Oversight and the Subcommittee on Science, Research and Technology of the Committee on Science and Technology.* 97th Congress, 1st Session, 8–9 June.

Kingsbury, David T. 1986. "Statement before the House of Representatives Subcommittee on Investigations and Oversight, and the Subcommittee on Natural Resources, Agricultural Research and Environment and the Subcommittee on Science, Research and Technology of the Committee on Science and Technology." *Coordinated Framework for the Regulation of Biotechnology: Hearings before the Subcommittee on Investigations and Oversight, and the Subcommittee on Natural Resources, Agricultural Research and Environment and the Subcommittee on Science, Research and Technology of the Committee on Science and Technology.* 99th Congress, 2nd Session, 23 July. No. 135.

Kloppenburg, Jack R. 1988. *First the Seed: The Political Economy of Plant Biotechnology, 1492–2000.* Cambridge: Cambridge University Press.

Kretchmer, Norman. 1991. "Why Not Have More Milk?" *Pediatrics* 88 (November): 1056–7.

Kronfeld, David S. 1989. "BST Milk Safety" (letter). *Journal of the American Veterinary Medicine Association* 195 (1 August): 288–9.

– 1991. "Safety of Bovine Growth Hormone (letter)." *Science* 251 (18 January): 256.

– 1993a. "Statement to the Food and Drug Administration Veterinary Medicine Advisory Committee (VMAC)." *Summary Minutes of VMAC Meeting.* Gaitersburg, MD, 31 March.

– 1993b. "Concerns about Bovine Somatropin (letter)." *Journal of the American Veterinary Medicine Association* 20 (15 July): 190–2.

– 1994. "Health Management of Dairy Herds Treated with Bovine Somatotropin." *Journal of the American Veterinary Medical Association* 204 (1 January): 116–30.

– 1996. "Statement to the Food and Drug Administration Veterinary Medicine Advisory Committee (VMAC)." Transcript of Proceedings to Review 28 Herd Post-Approval Monitoring Study and Findings from the Commercial Use of Sometribove. Rockville, MD, 20 November, 105–23.

Kuczynski, Alex. 1998. "Pursuing Potions: Fountain of Youth or Poisonous Fad?" *New York Times*, 14 April, B1, B9.

Kuhn, Thomas S. 1970. *The Structure of Scientific Revolutions*. 2d ed. Chicago and London: University of Chicago Press.

Kunkel, Mary Elizabeth. 1993. "Position of the American Dietetic Association: Biotechnology and the Future of Food." *Journal of the American Dietetic Association* 93 (February): 189–92.

Lambert, Eugene I. 1997. "The Reformation of Animal Drug Law: The Impact of 1996." *Food and Drug Law Journal* 52: 277–89.

Latour, Bruno, and Steve Woolgar. 1979. *Laboratory Life: The Social Construction of Scientific Facts*. Beverly Hills, CA, and London: Sage Publications.

Leary, Stephen. 1995. "Post-Modern Milk." *Toronto Star*, 19 February, C1, C8.

Leary, Warren. 1998. "Farm Animals Breeding Superbugs?" *Globe and Mail* 13 July, A7.

Leiss, William. 1990. "Problem Areas in the Management of Technology III: Managing the Risks and Consequences of Innovation." In *Managing Technology: Social Science Perspectives*, edited by Liora Salter and David Wolfe. Toronto: Garawood Press.

Leonard-Barton, Dorothy, and Gary Pisano. 1990. *Monsanto's March into Biotechnology (A)*. Discussion Paper 9-690-009. Cambridge, MA: Harvard Business School.

Lesser, William H., ed. *Animal Patents: The Legal, Economic and Social Issues*. New York: Stockton Press.

Levidow, Les. 1994. "Biotechnology Regulation as Symbolic Normalization." *Technology Analysis and Strategic Management* 6: 273–88.

Levidow, Les, and Susan Carr. 1996. "How Biotechnology Regulation Sets a Risk/Ethics Boundary." *Agriculture and Human Values* 14, no. 1 (March): 29–43.

"Life Science: A Dying Breed?" 2000. Agence France-Press, 20 November, from http://www.gene.ch/genet/2000/Nov/msg0053.html.

"Life Sciences – Green and Dying." 2000. *The Economist*, 16 November, at http://www.gene.ch/genet/2000/Nov/msg00053.html.

Longino, Helen E. 1990. *Science as Social Knowledge: Values and Objectivity in Scientific Inquiry*. Princeton, NJ: Princeton University Press.

Macklin, Arthur W. 1994. "Letter to the Prime Minister." 6 June. Toronto Food Policy Council Archives.

MacLeod, Robert. 1994. "Health Board Won't Swallow Hormone Plan." *Globe and Mail*, 24 March, A20.

Manley, John. 1997. "Statement. Speaking Notes to the House of Commons Standing Committee on Industry, Review of Bill C-91," Ottawa, 17 February.

– 1998. "Speaking Notes to the Ottawa Life Sciences Council National Conference and Exhibition," 17 November, at http://info.ic.gc.ca.

Marcus, Alfred A. 1988. "Risk, Uncertainty and Scientific Judgement." *Minerva* 26 (summer): 138–52.

Matas, Robert. 1995. "Ottawa to Let Citizens Sue Polluters." *Globe and Mail* 16 December, A1, A2.

Matte, Kempton L. 1994. "Statement before the House of Commons Standing Committee on Agriculture and Agri-Food." *Minutes of Proceedings and Evidence of the Standing Committee on Agriculture and Agri-Food Respecting Consideration of the Second Report of the Steering Committee, Pursuant to Standing Order 108(2), Consideration of Issues Relating to the Bovine Somatotropin Hormone (BST)*. 35th Parliament, 1st Session, No. 3. 8 March.

Matthews, Alexander S. 1998. "European Union Ban on Feed Additive Antibiotics Unjustified." Animal Health Institute, news release, 16 December.

McCarthy, Shawn. 1997. "Canada's Drug War." *Toronto Star* 13 March, D1, D4.

McGuire, M.A., and D.E. Bauman. 1996. "Regulation of Nutrient Use by Bovine Somatotropin: The Key to Animal Performance and Well-Being." Paper presented at the Ninth International Conference of Production Diseases in Farm Animals, Berlin.

McIlroy, Anne. 1998. "Tribunal Discusses Complaints over Drug-Approval Process." *Globe and Mail*, 24 December, A5.

McIlroy, Anne, and Laura Eggertson. 1997. "Health Branch Spared Cuts – For Now." *Toronto Star*, 24 September, A3.

McKenna, Barrie. 1994. "U.S. Discouraging Mandatory Food Labelling." *Globe and Mail*, 26 October, A4.

– 1996. "Dairy Products Prices to Rise 4 Percent." *Globe and Mail*, 31 July.

– 1997. "Ottawa Seeks Prescription for Drug Patent Battle." *Globe and Mail*, 17 February, B4.

McLaughlin, Gord. 1995. "Milk Meets Biotech." *Financial Post* 1 July.

McMahon, Eileen. 1995. "Nucleic Acid Sequences and Other Naturally Occurring Products: Are They Patentable in Canada?" Paper presented at Genetics and the Law Symposium, 21 October, Osgoode Hall, Toronto.

McMichael, Philip. 1992. "Tensions between National and International Control of the World Food Order: Contours of a New Food Regime." *Sociological Perspectives* 35, no. 2: 343–63.

McMichael, Philip and David Myhre. 1991. "Global Regulation vs. the Nation-state: Agro-Food Systems and the New Politics of Capital." *Capital and Class*, no. 43 (spring): 83–105.

McMurray, Scott. 1992. "Monsanto Plans to Write Down Assets, Cut Jobs." *Wall Street Journal*, 23 November, A3.

Melcher, Richard A., Amy Barrett, and Andrew Osterland. 1998. "Grains That Taste Like Meat?" *Business Week* 25 May, 44.

Mepham, T.B. 1992. "Public Health Implications of Bovine Somatotropin Use in Dairying: Discussion Paper." *Journal of the Royal Society of Medicine* 85 (December): 736–9.

Mepham, T.B., P.N. Schofield, W. Zumkeller, A.M. Cotterill. 1994. "Safety of Milk from Cows Treated with Bovine Somatotropin (letter)." *The Lancet* 344 (19 November): 1445–6.

Michaels, Adrian. 1999. "Monsanto, Pharmacia and Upjohn Shares Drop on Merger News." *Financial Post*, 21 December, C10.

"Milk with Experimental Hormone Is Safe for Humans, a Panel Finds." 1990. *New York Times*, 8 December, section 1, 14.

Miller, James P., and Ralph T. King Jr. 1995. "Monsanto to Get 49.9 Percent Stake in Calgene." *Wall Street Journal*, 29 June, B16.

Miller, Lee A. 1987. "Statement to the U.S. House of Representatives Subcommittee on Livestock, Dairy, and Poultry of the Committee on Agriculture." *Review of Status and Potential Impact of Bovine Growth Hormone: Hearing before the Subcommittee on Livestock, Dairy, and Poultry of the Committee on Agriculture*. 99th Congress, 2nd Session, Serial No. 99-51, 11 June. Washington, DC: U.S. Government Printing Office.

Miller, Margaret A. 1993. "Bovine Somatotropins: First Draft." *Toxicological evaluation of certain veterinary drug residues in food*. WHO Food Additives Series 31, prepared for the fortieth meeting of the Joint FAO/WHO Expert Committee on Food Additives (JECFA), Geneva, World Health Organization, 149–65.

Mills, Lisa. 1999. "Science and Social Context: The Regulation of Recombinant Bovine Growth Hormone (rbGH) in the United States and Canada, 1982–1998." PHD diss., Department of Political Science, University of Toronto.

Millstone, Erik, Eric Brunner, and Ian White. 1994. "Plagiarism or Protecting Public Health? (Commentary)." *Nature* 371 (20 October): 647–8.

Mitchell, G.A. 1993. "Statement to the Food and Drug Administration Veterinary Medicine Advisory Committee (VMAC)." *Summary Minutes of VMAC Meeting*. Gaitersburg, MD, 31 March.

Monsanto. 1993. "Official FDA Warning Label for Monsanto's Product Posi-

lac (Bovine Growth Hormone (BGH) or Bovine Somatotropin (BST)).” St Louis, MO.
– 1994a. “Posilac Bovine Somatotropin Enters U.S. Dairy Market.” *Monsanto Extra Report: A Special Supplement to Monsanto Company's Quarterly Report.* St Louis, MO: Monsanto.
– 1994b. *Technical Manual.* St Louis, MO: Monsanto.
– 1995. “BST Education Process Information Sheet.”
– 1996a. *Annual Report to Shareowners, 1995.* St Louis, MO: Monsanto.
– 1996b. *Monsanto Magazine.* No. 4
– 1997a. *Annual Report to Shareowners, 1996.* St Louis, MO: Monsanto.
– 1997b. *Monsanto Magazine.* No. 1.
– 1997c. Advertisement. *New York Times*, 19 August, A14, A15.
– 1998a. *Annual Report to Shareowners, 1997.* St Louis, MO: Monsanto.
– 1998b. “Status Update: Posilac® Bovine Somatotropin,” 6 August.
“Monsanto Told to Halt Promotion of Its Gene-Engineered Milk Drug.” 1991. *New York Times*, 13 February, B5.
“Monsanto's New Regime: Heavy Injections of Drugs and Biotechnology.” 1984. *Business Week*, 3 December, 64–6.
Moody's Industrial Manual. 1995. Vol. 2. New York: Moody's Investors Service.
Morrissey, Brian. 1995. “Standing Committee on Environment and Sustainable Development.” *Minutes of Proceedings and Evidence of the Standing Committee on Environment and Sustainable Development, Review of the Canadian Environmental Protection Act.* 30 January.
Morrow, David J., and Laura M. Holson. 1998. “Drug Giants Cancel Plans for a Merger.” *New York Times*, 24 February, C1, C8.
Mowling, Ray. 1995. “Letter to MPs,” 16 June. Toronto Food Policy Council Archives.
Myerson, Allen R. 1998. “Monsanto Settling Genetic Seed Complaints.” *New York Times*, 24 February, C2.
Naj, Amal Kumar. 1989. “Clouds Gather over the Biotech Industry.” *Wall Street Journal*, 30 January, B1, B5.
National Co-operative Research Act. 1984. Public Law 98-462, 96th Congress, 2d Session (11 October).
Nichols, Mark. 1995. “Messing with Milk.” *Maclean's Magazine*, 31 July, 24.
NIH (National Institutes of Health). 1991a. “Bovine Somatotropin and the Safety of Cows' Milk: National Institutes of Health Technology Assessment Conference Statement.” *Nutrition Review* 49 (August): 227–32.
– 1991b. “Special Communication: NIH Technology Assessment Conference Statement on Bovine Somatotropin.” *Journal of the American Medical Association* 255 (20 March): 1423–5.
– 1991c. “Technology Assessment Conference Statement: Bovine Somatotropin,” 5–7 December 1990 (draft), Bethesda, MD.

"NIH Panel Endorses Human Safety of BST." *Animal Pharm* 218 (December 21): 12.

National Research Council. Board of Agriculture: Committee on the Future of the Colleges of Agriculture in the Land Grant University System. 1996. *Colleges of Agriculture at the Land Grant Universities: Public Service and Public Policy*. Washington DC: National Academy Press.

North American Free Trade Agreement Secretariat. 1996. *Final Report of the Arbitral Panel Established Pursuant to Article 2008 in the Mattter of Tariffs Applied by Canada to Certain U.S.-Origin Agricultural Products, CDA-95-2008-01*, 2 December.

Northfield, Stephen. 1996. "Here's How to Play Those Hot Biotech Stocks." *Globe and Mail*, 16 November, B22.

O'Connor, Kevin W. 1989. "Congressional Perspectives." In *Animal Patents: The Legal, Economic and Social Issues*, edited by William H. Lesser. New York: Stockton Press.

OECD (Organization for Economic Cooperation and Development).1988a. *Biotechnology and the Changing Role of Government*. Paris, OECD.

– 1988b. *Report of a Group of Experts on the Social Aspects of New Technologies*. Paris, OECD.

– 1989. *Biotechnology: Economic and Wider Impacts*. Paris, OECD.

– 1992. *Biotechnology, Agriculture and Food*. Paris, OECD.

– 1996. *Science, Technology and Industry Outlook*. Paris, OECD.

Office of the Inspector General. Department of Health and Human Services. 1991. *Need for the Food and Drug Administration to Review Possible Improper Pre-Approval Promotional Activities*. A-15-91-00007. Richard P. Kusserow, Inspector General. Washington, DC: Inspector General.

– 1994. *Follow-Up Review of Possible Improper Pre-Approval Promotion Activities*. A-15-93-00018. June Gibbs-Brown, Inspector-General, September, Washington, D.C.

Olanrewaju, Hammed A., Eric. D. Sanzenbacher, and Edward R. Seidel. 1996. Insulin-Like Growth Factor I in Suckling Rate Gastric Contents. *Digestive Diseases and Sciences* 41 (July): 1392–7.

Olanrewaju, Hammed, Laju Patel, and Edward R. Seidel. 1992. "Trophic Action of Local Intraileal Infusion of Insulin-Like Growth Factor I: Polyamine Dependence." *American Journal of Physiology, Endocrinology and Metabolism* 26 (August): E282–6.

OMB (Office of Management and Budget). 1994. *Use of Bovine Somatotropin (BST) in the United States: Its Potential Effects - A Study Conducted by the Executive Branch of the Federal Government*. Washington, DC: OMB.

Oosterhoff, Peter. 1994. "Statement before the House of Commons Standing Committee on Agriculture and Agri-Food." *Minutes of Proceedings and Evidence of the Standing Committee on Agriculture and Agri-Food Respecting Consideration of the Second Report of the Steering Committee, Pursuant to*

Standing Order 108(2), Consideration of Issues Relating to the Bovine Somatotropin Hormone (BST). 35th Parliament, 1st Session, No. 3. 7 March.

OSTP (Office of Science and Technology Policy). 1986. "Coordinated Framework for the Regulation of Biotechnology". *Federal Register*, 51, no. 123 (26 June): 23302.

– 1992. "Exercise of Federal Oversight within Scope of Statutory Authority: Planned Introductions of Biotechnology Products into the Environment, Announcement of Policy." *Federal Register*, 57, no. 39 (27 February): 6753.

OTA (Office of Technology Assessment). 1986. *Technology, Public Policy, and the Changing Structure of American Agriculture*. Washington, DC: U.S. Government Printing Office.

– 1988. *New Developments in Biotechnology*. Vol. 4, *U.S. Investment in Biotechnology*. Special Report OTA-BA-360. Washington, DC: U.S. Government Printing Office.

– 1990. *New Developments in Biotechnology: Patenting Life*. New York and Basel: Marcel Dekker.

– 1991a. *Biotechnology in a Global Economy*. OTA-BA-494 Washington, DC: U.S. Government Printing Office.

– 1991b. *U.S. Dairy Industry at a Crossroad: Biotechnology and Policy Choices*. Special Report, OTA-F-470, Washington, DC: U.S. Government Printing Office.

Palmer, Jay. 1998. "New DuPont." *Barron's*, 11 May, 31–6.

Patent and Trademark Amendments Act of 1980, PL 96-517, H.R. 6933, 96th Congress, 2nd Session, 12 December.

Pear, Robert. 1991. "Panel Calls Federal Drug Agency Unable to Cope with Rising Tasks." *New York Times*, 11 April, A1, B11.

Pearce, Deborah. 1994. "Uncowed Opponents Consider Udder Side of Synthetic Hormone." *Times Colonist*, 7 August.

Peel, Colin J. and Bauman, D.E. 1987. "Somatotropin and Lactation." *Journal of Dairy Science* 70: 474–86.

Pell, Alice N., D.S. Tsang, B.A. Howlett, M.T. Huyler, V.K. Meserole, W.A. Samuels, G.F. Hartnell, and R.L. Hintz. 1992. "Effects of a Prolonged Release Formulation of Sometribove (n-Methionyl Bovine Somatotropin) on Jersey Cows." *Journal of Dairy Science* 75: 3416–31.

Pilon, Lise. 1994. "Statement before the House of Commons Standing Committee on Agriculture and Agri-Food." *Minutes of Proceedings and Evidence of the Standing Committee on Agriculture and Agri-Food Respecting Consideration of the Second Report of the Steering Committee, Pursuant to Standing Order 108(2), Consideration of Issues Relating to the Bovine Somatotropin Hormone (BST)*. 35th Parliament, 1st Session, No. 3. 8 March.

Plein, L. Christopher. 1991. "Popularizaing Biotechnology: The Influence of Issue Definition." *Science, Technology and Human Values* 16 (autumn): 474–90.

Pollack, Andrew. 1998. "Venture Capital for an Orphan: Agricultural Biotechnology." *New York Times*, 14 April, C12.

Pollak, Michael N., James F. Perdue, Richard G. Margolese, Kathy Baer, and Martine Richard. 1987. "Presence of Somatomedin Receptors on Primary Human Breast and Colon Carcinomas." *Cancer Letters* 38: 223–30.

Pollak, Michael, Constantin Polychronakos, and Harvey Guyda. 1989. "Somatostatin Analogue SMS 201-995 Reduces Serum IGF-I Levels in Patients with Neoplasms Potentially Dependent on IGF-I." *Anticancer Research* 9: 889–92.

Pollak, Michael, Constantin Polychronakos, and Martine Richard. 1990. "Insulinlike Growth Factor I: A Potent Mitogen for Human Osteogenic Sarcoma." *Journal of the National Cancer Institute* 82 (February 21): 301–5.

Pollak, Michael, Joseph Costantino, Constantin Polychronakos, Sue-Ann Blauer, Harvey Guyda, Carol Redmond, Bernard Fisher, and Richard Margolese. 1990. "Effect of Tamoxifen on Serum Insulinlike Growth Factor I Levels in Stage I Breast Cancer Patients." *Journal of the National Cancer Institute* 82: 1693–7.

Powell, Douglas, and William Leiss. 1997. *Mad Cows and Mother's Milk: The Perils of Poor Risk Communication*. Montreal and Kingston: McGill-Queen's University Press.

"Press Council Dismisses Complaint." 1996. *Globe and Mail*, 6 November, A11.

Prosser, C.G., I.R. Fleet, I.C. Hart, and R.B. Heap. 1987. "Changes in Concentrations of IGF-I in Milk during BGH Treatment in the Goat." *Journal of Endocrinology* 112 (March supplement): abstract 65.

Rabinow, Paul. 1992. "Severing the Ties: Fragmentation and Dignity in Late Modernity." In *Knowledge and Society: The Anthropology of Science and Technology*. Vol. 9, edited by Arie Rip, David J. Hess, and Linda L. Layne. Greenwich, CT, and London: JAI Press.

RAFI (Rural Advancement Fund International). 1997. "The Life Industry 1997: The Global Enterprises that Dominate Commercial Agriculture, Food and Health." *RAFI Communique*: November/December.

– 1998a. "Seed Industry Consolidation: Who Owns Whom?" *RAFI Communique*: July/August.

– 1998b. "The Terminator Technology: New Genetic Technology Aims to Prevent Farmers from Saving Seed." *RAFI Communique*: March/April.

– 1999. "What Do You Get When You Make a GM Cross of Pharmacia-Upjohn and Monsanto?" *Pharma-geddon*, 21 December, at http://www.gene.ch/genet/1999/Dec/msg00030.html.

– 2000. "Canadian Court Rules Mammals Can Be a Patented Invention." *Financial Express*, 14 August, at wysiwyg://5/http://www.expressindia.com/fe/daily/20000814/fc014039.html.

Ramirez, Anthony. 1996. "High-Tech Milk in Schools Worrying Some Parents." *New York Times*, 3 March, section 13, 6.

Raub, William. 1981. "Statement to the Subcommittee on Investigations and Oversight and the Subcommittee on Science, Research and Technology of the Committee on Science and Technology, U.S. House of Representatives. *Commercialization of Academic Research: Hearings before the Subcommittee on Investigations and Oversight and the Subcommittee on Science, Research and Technology of the Committee on Science and Technology.* 97th Congress, 1st Session, 8–9 June.

rBST Internal Review Team. Health Protection Branch, Health Canada. 1998. "rBST (Nutrilac) 'Gaps Analysis' Report." 21 April, at http://www.nfu.ca.

Reish, Marc S. 1995. "Monsanto Continues to Shift Away from High-Volume Commodities." *Chemical and Engineering News*, 6, 13 & 14 November.

"Research Reports: De Kalb Genetics." 1997. *Barron's*, 27 January, 48.

Reuters News Agency. 1998a. "Tamoxifen Needs More Research, British Say." *Globe and Mail*, 8 April, A6.

– 1998b. "Monsanto to Buy Cargill Seed Units." *Globe and Mail*, 30 June, B13.

Richards, Bill. 1989. "Dairy Farmers, Drug Firms Clash Over Cow Hormone." *Wall Street Journal*, 4 May, B1, B11.

Rifkin, Jeremy. 1987. "Statement to the U.S. House of Representatives Subcommittee on Livestock, Dairy, and Poultry of the Committee on Agriculture." *Review of Status and Potential Impact of Bovine Growth Hormone: Hearing before the Subcommittee on Livestock, Dairy, and Poultry of the Committee on Agriculture.* 99th Congress, 2nd Session, Serial No. 99-51, 11 June. Washington, DC: U.S. Government Printing Office.

Ritter, Leonard. 1994a. "Letter to Marbeth Greer," 26 April. Toronto Food Policy Council Archives.

– 1994b. "Statement before the House of Commons Standing Committee on Agriculture and Agri-Food." *Minutes and Proceedings* (8 March).

Roberts, Edward R. 1987. "Statement to the U.S. House of Representatives Subcommittee on Livestock, Dairy, and Poultry of the Committee on Agriculture." *Review of Status and Potential Impact of Bovine Growth Hormone: Hearing before the Subcommittee on Livestock, Dairy, and Poultry of the Committee on Agriculture.* 99th Congress, 2nd Session, Serial No. 99-51, 11 June. Washington, DC: U.S. Government Printing Office.

Rogers, Arthur. 1997a. "Euro Food Safety Chief Says 'Expect No Miracles.'" *The Lancet* 349 (24 May): 1529.

– 1997b. "European Science Committee Proposed." *The Lancet* 349 (21 June): 1823.

Rogers, Karen Keeler. 1996. "Fields of Promise: Monsanto and the Development of Agricultural Biotechnology – Setting the Course." *Monsanto Magazine*, no. 4.

– 1997. "Fields of Promise: Monsanto and the Development of Agricultural Biotechnology – Creating the Future." *Monsanto Magazine*, no. 1.

Rotman, David. 1997. "Monsanto Spends $1.02 Billion for Corn Seed Business." *Chemical Week*, 15 Janaury, 7.

Royal College of Physicians and Surgeons of Canada. Expert Panel on Human Safety of rbST. 1999. *Report Prepared for Health Canada*. January.

Royal Society of Canada. 2001. *Elements of Precaution: Recommendations for the Regulation of Food Biotechnology in Canada, An Expert Panel Report on the Future of Food Biotechnology*. Prepared at the request of Health Canada, the Canadian Food Inspection Agency, and Environment Canada. Ottawa, Ontario, January.

Ruivenkamp, Guido. 1988. "Emerging Patterns in the Global Food Chain." In *Biotechnology Revolution and the Third World*. New Delhi: Research and Information Systems for the Non-Aligned and Other Developing Countries.

Rural Vermont. 1996. "Vermont Attorney General Gives up Fight for Blue Dots and Consumer's Right to Know." News release. 30 August.

– 1997a. "Legislative Update: rbGH." *Rural Vermont Report* 8 (spring): 1–2.

– 1997b. "Rural Vermont Hails Ben & Jerry's rbGH Victory." News release. 14 August.

Salter, Liora.1988. *Mandated Science: Science and Scientists in the Making of Standards*. Dordrecht, Boston, and London: Kluwer Academic Publishers.

Sanders, Bernard. 1993. "Statement to the Food and Drug Administration Veterinary Medicine Advisory Committee (VMAC)." *Summary Minutes of VMAC Meeting*. Gaitersburg, MD, 31 March.

Sara, Vicki R., and Kerstin Hall. 1990. "Insulin-Like Growth Factors and Their Binding Proteins." *Physiological Reviews* 70 (July): 591–607.

Saunders, Doug. 1995a. "Milk Hormone Battle about to Heat Up." *Globe and Mail* 22 June, A10.

– 1995b. "Drug Firms Upset about Expected Ban: Monsanto, Eli Lilly Threaten to Pull Some Research Investments out of Canada over Cow Hormone." *Globe and Mail*, 27 June, B7.

– 1995c. "Milk's Raging Hormone." *Globe and Mail* 13 July, A13.

Savitz, Eric J. 1996. "Unappreciated: A Money Manager Sees Lots of Bargains among the Beaten-Down Biotechs." *Barron's*, 2 December, 22.

Schneider, Keith. 1988. "Biotechnology's Cash Cow." *New York Times Magazine*, 12 June, 44–53.

– 1989a. "Gene-Altered Farm Drug Starts Battle in Milk States." *New York Times*, 29 April, section 1, 1, 8.

– 1989b. "Vermont Resists Some Progress in Dairying." *New York Times*, 27 August, E4.

– 1990a. "FDA Accused of Improper Ties in Review of Drug for Milk Cows." *New York Times*, 12 January, A21.

– 1990b. "Consumer Group Questions Milk Hormone's Safety." *New York Times*, 4 December, A29.
– 1993. "U.S. Approves Use of Drug to Raise Milk Production." *New York Times*, 6 November, section 1, 1 & 9.
– 1994a. "Grocers Challenge Use of New Drug for Milk Output." *New York Times*, 4 February, A1, A14.
– 1994b. "Question Is Raised on Hormone Maker's Ties to FDA Aides." *New York Times*, 18 April, A9.
– 1994c. "Despite Critics, Dairy Farmers Increase Use of a Growth Hormone in Cows." *New York Times*, 30 October, S1, 11.
Schneider, N.M. 1981. "Scrambling on the Gene Train." *Genetic Engineering News*, January/February, 4.
Schneiderman, Howard A. 1982. "Testimony before the U.S. House of Representatives Subcommittee on Investigations and Oversight and the Subcommittee on Science, Research and Technology of the Committee on Science and Technology." *University/Industry Cooperation in Biotechnology: Hearings before the Subcommittee on Investigations and Oversight and the Subcommittee on Science, Research and Technology*, Ninety-seventh Congress, Second Session, 16, 17 June, Washington, DC: U.S. Government Printing Office.
Schwartz, John. 1994. "Probe of 3 FDA Officials Sought." *Washington Post*, 19 April, A3.
Scoffield, Heather. 1995. "Letter to Shareowners." *Monsanto Annual Report to Shareowners*. St Louis, Missouri.
– 1996. "Letter to Shareowners." *Monsanto Annual Report to Shareowners*. St Louis, Missouri.
– 1997. "A New Era of Value Creation." *Monsanto Annual Report to Shareowners*. St Louis, Missouri.
– 1998. "Cuts Put Biotech on Critical List." *Globe and Mail*, 15 January, B1, B6.
– 2000. "Canada Loses Patent Case, Higher Drug Prices Feared." *Globe and Mail*, 4 March, A1, A23.
Scientific Committee on Animal Health and Animal Welfare. 1999. *Report on Animal Welfare Aspects of the Use of Bovine Somatotrophin*. European Commission, DG24, 10 March.
Scientific Committee on Veterinary Measures Relating to Public Health. 1999. Report on Public Health Aspects of the Use of Bovine Somatotrophin. European Commission, DG24, 15–16 March, at http://europa.eu.int/comm/dg24/health/sc/scu/out19_en.html.
Shapiro, Bob. 1996. "Letter to Our Shareowners." *1995 Monsanto Annual Report*, St Louis, MO.
Siler, Julia Flynn, and John Carey. 1991. "Is Monsanto "Burning Money" in its Biotech Barn?" *Business Week*, 2 September, 74–5.

Skelton, W. Douglas. 1993. "Statement of the American Medical Association to the Food and Drug Administration Veterinary Medicine Advisory Committee (VMAC). *Summary Minutes of VMAC Meeting*. Gaitersburg, MD. 31 March.

Skogstad, Grace. 1987. *The Politics of Agricultural Policy-Making in Canada*. Toronto, Buffalo, and London: University of Toronto Press.

Slaughter, Sheila, and Gary Rhoades. 1996. "The Emergence of a Competitiveness R&D Policy Coalition and the Commercialization of Academic Science and Technology." *Science, Technology and Human Values* 21 (summer): 303–39.

Smith, K.L., R.J. Eberhart, R.J. Harnon, D.E. Jasper, S.C. Nickerson, and G.E. Shook. 1988. *Protocol for the Evaluation of Mastitis in Efficacy Studies of Bovine Somatotropin and Production Drugs in Dairy Cattle, Addendum to a Technical Assistance Document for Efficacy Studies of Bovine Somatotropin (BST) in Lactating Dairy Cows*, 17 March, 1–9.

Soil Association. 1998. "Failure to Control Farm Antibiotics Poses Serious Threat to Human Health." News release, 23 April, at http://www.soilassociation.org/SA/SAWeb.nsf.

Specter, Michael. 1998. "Europeans Revolt against Test-Tube-Altered Future." *Globe and Mail*, 21 July, A11.

St Louis Post-Dispatch. 1992. 25 February, 1.

Stayer, Robert. 1995. "Monsanto Offers Discounts to Dairy Farmers." *St Louis Post-Dispatch*, 22 October, 1, 8.

Stecklow, Steve. 1999. "How a U.S. Gadfly and a Green Activist Started a Food Fight." *Wall Street Journal*, 20 November, A1, A10.

Stemler, Walter E. 1987. "Statement to the U.S. House of Representatives Subcommittee on Livestock, Dairy, and Poultry of the Committee on Agriculture." *Review of Status and Potential Impact of Bovine Growth Hormone: Hearing before the Subcommittee on Livestock, Dairy, and Poultry of the Committee on Agriculture*. 99th Congress, 2nd Session, Serial No. 99-51, 11 June. Washington, DC: U.S. Government Printing Office.

Stennes, B.K., R.R. Barichello, J.D. Graham.1990. "Bovine Somatotropin and the Canadian Dairy Industry: An Economic Analysis." Working paper 1/91, Department of Agriculture, Ottawa.

Stevenson-Wydler Technology Innovation Act of 1980, Public Law 96-480, Amended 1986 PL 99-502.

Stewart, Claire H., and Peter Rotwein. 1996. "Growth, Differentiation, and Survival: Multiple Physiological Functions for Insulin-Like Growth Factors." *Physiological Reviews* 76 (October): 1005–1026.

Stonehouse, D. Peter. 1987. "A Profile of the Canadian Dairy Industry and Government Policies." Working paper 4/87, Department of Agriculture, Ottawa, March.

Stoneman, Don. 1995. "Health Canada Demands More on BST." *Farm and Country*, 10 October, A1.

Szkotnicki, Jean. 1994. "Statement before the House of Commons Standing Committee on Agriculture and Agri-Food." *Minutes and Proceedings*, 8 March.

Tabuns, Peter. 1995. "Draft Letter to Bob Speller," 23 August.

Taylor, Paul. 1998. "Cancer Cases, Costs to Climb." *Globe and Mail*, 8 April, A1, A6.

Thayer, Ann M. 1996. "Market, Investor Attitudes Challenge Developers of Biopharmaceuticals." *Chemical and Engineering News*, 12 August, 13–21.

Thomas, J.W., R.A.Erdman, D.M. Galton, R.C. Lamb, M.J. Arambel, J.D.Olson, K.S. Masden, W.A.Samuels, C.J. Peel, and G.A. Green. 1991. "Responses by Lactating Cows in Commercial Dairy Herds to Recombinant Bovine Somatotropin." *Journal of Dairy Science* 74: 945–64.

Thomson Wire Service. 1994. "Controversial Dairy Drug on Hold." *Winnipeg Free Press*, 18 August.

Tolchin, Martin. 1990. "Gene-Altered Item Approved by FDA." *New York Times*, 25 March, A27.

Toronto Food Policy Council (TFPC). 1994a. "Toronto Food Policy Council Urges Federal Government Not to Licence Recombinant Bovine Growth Hormone (rBGH) and to Reform Federal Drug Review Process." *Press Release*. March 23.

– 1994b. "Draft Letter to Agriculture Minister Ralph Goodale," August. Toronto Food Policy Council Archives.

– 1995. "Setting a New Direction: Changing Canada's Agricultural Policy-Making Process." Discussion paper no. 4.

"The Ultimate Synthesizer? Photosynthesis, of Course." 1995. *Barron's*, 21 August, 10.

United States. House of Representatives. Subcommittee on Investigations and Oversight, and the Subcommittee on Science, Research, and Technology of the Committee on Science and Technology. 1981. *Commercialization of Academic Biomedical Research: Hearings Before the Subcommittee on Investigations and Oversight, and the Subcommittee on Science, Research, and Technology of the Committee on Science and Technology*. 97th Congress, 1st Session, 8, 9 June.

– Subcommittee on Livestock, Dairy, and Poultry of the Committee on Agriculture. 1986. *Status and Review of Bovine Growth Hormone, Hearing before the Subcommittee on Livestock, Dairy, and Poultry of the Committee on Agriculture*. 99th Congress, Second Session, 11 June, Serial No. 99-51, Washington, DC: U.S. Government Printing Office.

– Subcommittee on Courts, Civil Liberties and the Administration of Justice and the Committee of the Justice Department. 1987. *Patents and the Constitution: Transgenic Animals, Hearing before the Subcommittee on Courts,*

Civil Liberties and the Administration of Justice and the Committee of the Justice Department. 100th Congress, 1st Session, No. 23. 11 June, 22 July, 21 August, and 5 November.

United States. Senate. 1962. Drug Amendments of 1962: Senate Report No. 1744, 19 July, 1962 to accompany s. 1552, in *United States Code Congressional and Administrative News*. 87th Congress, Second Session, vol. 2.

Van Wyk, Judson J., Kerstin Hall, J. Leo Van Den Brande, Robert P. Weaver, Knut Uthne, Raymond L. Hintz, John H. Harrison, and Paul Mathewson. 1972. "Partial Purification from Human Plasma of a Small Peptide with Sulfation Factor and Thymidine Factor Activities." In *Growth and Growth Hormone: Proceedings of the Second International Symposium on Growth Hormone*, edited by A. Pecile and E.E. Muller, Milan, 5–7 May, 1971. Amsterdam: Excerpta Medica.

Vidal, John. 1999. "How Monsanto's Mind was Changed." *Guardian*, 9 October.

VMAC (Veterinary Medicine Advisory Committee of the Food and Drug Administration). 1993. *Summary Minutes of Meeting*. Gaitersburg, MD, 31 March.

– 1996. "Transcript of Proceedings to Review 28 Herd Post-Approval Monitoring Study and Findings from the Commercial Use of Sometribove," Rockville, MD, 20 November.

Wade, Robert. 1996. "Globalization and Its Limits: Reports of the Death of the National Economy are Greatly Exaggerated." In *National Diversity and Global Capitalism*, edited by Suzanne Berger and Ronald Dore. Ithaca and London, Cornell University Press.

Ward, Neil, and Reidar Almas. 1997. "Explaining Change in the International Agro-Food System." *Review of International Political Economy* 4, no. 4 (winter): 611–29.

Weldon, Virginia W. 1993. "Mastitis and Antibiotics. Statement to the Food and Drug Administration's Veterinary Medicine Advisory Committee (VMAC)." *Summary Minutes of VMAC Meeting*. Gaithersburg, MD. 31 March.

White, T.C., K.S. Masden, R.L. Hintz, R.H. Sorbet, R.J. Collier, D.L. Hard, G.F. Hartnell, W.A. Samuels, G. de Kerchove, F. Adriaens, N. Craven, D.E. Bauman, G. Bertrand, Ph. Bruneau, G.O. Gravert, H.H. Head, J. T. Huber, R.C. Lamb, C. Palmer, A.N. Pell, R. Phipps, R.Weller, G. Piva, Y. Rijpkema, J. Skarda, F. Vedeau, and C. Wollny. 1994. "Clinical Mastitis in Cows Treated with Sometribove (Recombinant Bovine Somatotropin) and Its Relationship to Milk Yield." *Journal of Dairy Science* 77: 2249–60.

Wiktorowicz, Mary E. 2000. "Shifting Priorities at the Health Protection Branch: Challenges to the Regulatory Process." *Canadian Public Administration* 43, no. 1 (spring): 1–22.

Williamson, Alistair D., and Dorothy Leonard-Barton. 1991. *Monsanto's March into Biotechnology (B)*. Discussion Paper 9-692-066. Harvard Business School, Cambridge, MA.

Wilson, Ewen M. 1987. "Statement to the U.S. House of Representatives Subcommittee on Livestock, Dairy, and Poultry of the Committee on Agriculture." *Review of Status and Potential Impact of Bovine Growth Hormone: Hearing before the Subcommittee on Livestock, Dairy, and Poultry of the Committee on Agriculture.* 99th Congress, 2nd Session, Serial No. 99-51, 11 June. Washington, DC: U.S. Government Printing Office.

Wilson, Lloyd L. 1993. "Concerns about Bovine Somatotropin (letter)." *Journal of the American Veterinary Medicine Association* 20 (15 July): 192.

Winkelaar, Susan. 1995. "Dairy Producers Avoiding Hormone." *Winnipeg Free Press*, 4 February.

Wirth, David A. 1994. "The Role of Science in the Uruguay Round and NAFTA Trade Disciplines." *Cornell International Law Journal* 27: 818–58.

"Wisconsin Leaders Back Curb on Sale of Growth Hormone." *New York Times*, 23 March, A12.

"World Trade Organization Denies Canada's Appeal for Shorter Patent Protection." 2000. Canada NewsWire, at http://www.newsire.ca/releases/September2000/19/c4903.html.

Wright, Susan. 1991. "The Social Warp of Science: Writing the History of Genetic Engineering Policy." *Science, Technology and Human Values* 16 (autumn): 79–101.

– 1994. *Molecular Politics: Developing American and British Regulatory Policy for Genetic Engineering, 1972–1982.* Chicago and London: University of Chicago Press.

Wyatt, Edward. 1996. "Corporate America is Courting Agricultural Biotech." *New York Times*, 3 March, F11.

WTO (World Trade Organisation). 1994. "The WTO Agreement on the Application of Sanitary and Phyto-Sanitary Measures (SPS Agreement)," at http://www.wto.org/english/tratop_e/sps_e/spsagr_ e/htm.

– 1998. "Understanding the Agreement on Sanitary and PhytoSanitary Measures," May, at http://www.wto.org/english/tratop_e/sps_e/spsund_e.htm.

Wynne, Brian. 1996. "May the Sheep Safely Graze? A Reflexive View of the Expert-Lay Knowledge Divide." In *Risk, Environment, and Modernity: Towards a New Ecology*, edited by Scott Lash, Bronislaw Szerszynski, and Brian Wynne. London: Thousand Oaks; New Delhi: Sage Publications.

Xian, C.J., C.A. Shoubridge, and L.C. Read. 1995. "Degradation of IGF-I in the Adult Rat Gastrointestinal Tract is Limited by a Specific Antiserum or the Dietary Protein Casein." *Journal of Endocrinology* 146 (August): 215–25.

Yakaubski, Konrad, and Stephen Northfield. 1996. "The Biotech Bazaar." *Globe and Mail*, 27 July, B1, B3.

Young, Frank E. 1987. "Statement to the U.S. House of Representatives Subcommittee on Livestock, Dairy, and Poultry of the Committee on Agriculture." *Review of Status and Potential Impact of Bovine Growth Hormone:*

Hearing before the Subcommittee on Livestock, Dairy, and Poultry of the Committee on Agriculture. 99th Congress, 2nd Session, Serial No. 99-51, 11 June. Washington, DC: U.S. Government Printing Office.

Young, Ian. 1997. "Rhone-Poulenc and Merck Merge Units." *Chemical Week*, 1–8 January, 14.

Yoxen, Edward. 1983. *The Gene Business: Who Should Control Biotechnology?* London and Sydney: Pan Books.

Index